Landscape You Can Eat

LANDSCAPE YOU CAN EAT

Allan A. Swenson

DAVID McKAY COMPANY, INC.
New York

Library of Congress Cataloging in Publication Data

Swenson, Allan A
Landscape you can eat.

Includes index.
1. Fruit-culture. 2. Organic gardening.
3. Landscape gardening. I. Title.
SB357.24.S93 634 77–10491
ISBN 0–679–50647–0
ISBN 0–679–50669–1 pbk.

10 9 8 7 6 5 4 3 2 1

Book design: H. Roberts

To Philip Alampi, who saw and cultivated the talents of a budding young gardener, and with proper pruning helped shape a growing career. For that early guidance, encouragement, and challenge, continuing thanks.

Contents

Acknowledgments

During the many years that my family and I raised fruit and berries around our various homes, we were actually designing our own fruitful landscapes. It seemed easy and natural to plant a few fruit trees each year, plus some berry bushes here and there. When it came to putting all those years into one book, I realized how many people had helped me over the hurdles in those years gone by: from my professors at Rutgers University, who suffered through my naive questions, to nurserymen, who freely offered valuable growing tips. Dozens, perhaps hundreds of people, have really had a part in the preparation of this book.

To all of them, especially the dedicated men and women at the fruit breeding and testing stations, and those with the Geneva Experiment Station in particular, I am indebted. Your efforts have helped make gardens as well as farms across America more bountiful. All of us who enjoy the fruits of our harvests each year appreciate your efforts. May your work in the future prove as abundantly fruitful for generations to come.

Introduction

When winter winds howl and your garden lies barren outside the window, there's nothing tastier for starting the day than hot toast with homemade jam. Then, when your day's activities are over and homemade apple pie from your own apples is steaming on the table, that's the perfect finale for the day.

Perhaps some of the greatest joys of good gardening are in the delicious eating you can enjoy from the previous season's fruitful landscape.

Talking with friends and neighbors over the past few years, I've noticed a resurgence in the desire to make gardens and home grounds more productive. There's an urge in the air these days to be more self-sufficient, more capable of growing a wider range of plants, even in small backyard plots.

During hundreds of conversations, people have told me that they would like to enjoy more fruit trees and berry bushes, but they just didn't have the room. I was enjoying some good old-fashioned jam one wintery Sunday afternoon when it occurred to me that there is indeed a way to enjoy the fruits of garden labors even on small plots. That's the premise of this book: you *can* have an abundantly fruitful home landscape.

During the past two years, I've interviewed many good growing friends around this country. I've talked with plant breeders, pomologists who know their fruit business, farmers and orchardists, and legions of practical home gardeners.

On one point we all agree. Fruit growing isn't as difficult as most people believe. In fact, fruit trees can be successfully grown in many parts of the country. Berry bushes, too, thrive across America. For years, many of us seem to have thought that fruit growing requires lots of space and special talents. That's really not so. With a little extra effort at planting time and some knowledge of pruning, tending, and caring for fruit trees and berry bushes, any family can enjoy a more fruitful living from its land.

Fact is, fruit trees and berry bushes are versatile. They fit well into outdoor landscapes. They provide shade and appealing shapes and forms to grace our homes, gardens, lawns, and property lines. They can grow by themselves or in groups. Berry bushes have a variety of shapes and growth patterns that fit well into landscape plantings.

It seems logical to take advantage of the extra pleasure, and the bonuses, that fruit trees, grapevines, and berry bushes offer us. Besides having value as decorative plants, they also can reward us with tastier eating. With those thoughts in mind, and with the help of hundreds of good growing friends and fruit-growing experts, too, I have written this book. It is somewhat different from most of my others, since it isn't merely a gardening how-to book. It is designed to stimulate your appetite for more flavorful living as well as help you expand your growing horizons to enjoy the fruits of your gardening activities even more.

Landscape You Can Eat is dedicated to all of you who want to enjoy more rewarding, more productive lives on your own plot here on the good planet Earth. Multipurpose landscaping, using fruit trees and berry plants as part of your total landscaping, will provide that extra dimension in good eating. The trees and bushes you grow are lovely to look at—food for the soul. The fruits you harvest provide those tasty rewards for the body: fresh fruit, pies, jams, jellies, preserves, and the hundreds of delights you can make when you adopt a multipurpose landscape plan.

On our home grounds previously and here Down East today, we're replanting and rearranging our gardening efforts toward more fruitful productivity. It's fun, tasty, and worthwhile. Having lived in other regions, and traveling every year to visit many parts of America, I've pooled the best information from many sources in this book. Each chapter is also based on personal growing experiences, from the time way back to my youngest days on the farm tending acres of fruit trees to smaller home plantings in recent years.

I hope that you enjoy this guide to tastier living. Each tree you plant, each bush and shrub that bears some fruit, will bring you that much closer to the more flavorful, abundant life from *Landscape You Can Eat.*

Allan A. Swenson
Windrows Farm
Kennebunk, Maine
1977

1. Plan Your Fruitful Plantscape

You can enjoy lots more delicious eating when you landscape fruitfully. Actually, you can even eat your own landscape when you put multipurpose fruit trees and berry bushes together or use them in a variety of ways around your home grounds. Especially on small plots where space is severely limited, fruit trees and berry bushes offer greater versatility than shade trees or decorative hedges.

As you plan new landscape plantings or renovations of your present outdoor living areas, think fruitfully. Often a berry bush or two can be substituted for another type of shrub. Privet hedges are attractive and have their place as more formal plantings. However, a hedge of blueberries can be nearly as dense plus yield its bounty every year. In fall, unlike the privet and other types of hedges that merely drop their leaves, blueberries provide a red-to-scarlet foliage display for you.

Spirea, bridal wreath, and similar lovely flowering shrubs unfurl their own distinctive blossoms every spring for glorious displays. Enjoy them as you will. They add a bright spot to your home grounds. However, don't overlook some of the tasty berries that can grace a corner, frame a doorway, or line a pathway. Currants bloom well, though not quite so spectacularly as other flowering shrubs. But when their tastier advantages are considered, perhaps a few deserve a place amid showier shrubs.

There is real value in blending different forms, shapes, and leaf patterns as part of a total outdoor scene. When you select furniture for a room, you look for pieces that complement each other but aren't necessarily exactly alike. Some standout accent pieces always lend a special touch. The same is true with outdoor living rooms.

As you design your property borders, consider fruitful bushes. Hedges are helpful to mark your backyard lines, screen an unwanted view, or partition off a children's play area from an outdoor entertaining area. If you have a long stretch of wall, try a fruitful hedge along it. The more natural growth pattern of living hedges will break up the stark lines of long walls. A tree or two trained in espalier form also can dramatically enhance a large wall.

Long hedges alone can be striking and eye-catching parts of your plantings. Sometimes straight lines look good in more formal designs. At times, weaving or bending the planting pattern adds an extra touch of grace to the overall picture. Blueberries lend themselves to a more tidily trained hedge effect than raspberries or blackberries. They can be trained into graceful hedgerow plantings to block out undesired views of a neighbor's land, hide a compost pile or two, or separate one part of your yard from another.

Currant bushes in a row can be trained to a quasi hedge effect; but they're more natural in rows of rounder, bushier plants. Even set alone, they perform their fruitful function here and there among other plants in beds and borders, along a fence, or in a group where they add their own distinctive shapes to your plantscapes.

There's no rule that says fruit bushes must be planted in one spot, flowers in another, vegetables elsewhere. Masses of flowers planted in their own beds or borders admittedly have striking and beautiful effects. Vegetables may be more conveniently restricted to their own patch, it's true. However, European gardeners who have small plots find that mix-and-match is a lovely and more productive way to achieve the most abundance from their land. New interest in this logical idea among American gardeners has increased these past few years, and for good growing reasons.

It has been said you can't eat your cake and have it too. That's true. But with multipurpose landscaping, you can enjoy the blooms of bushes and trees each spring, their fruits in season, and foliage in the fall. What's more, you can taste the difference from a fruitful landscape as you and your family pick and eat the apples, peaches, pears, and plums, the raspberries, blackberries, blueberries, and mounds of strawberries, too. In addition to the fresh taste of these harvests, you have other options to really stretch that fruitful landscape to year-round enjoyment. Pies and cakes, preserves, jams and jellies, canned and frozen fruits and berries, plus wine and brandy are all extra benefits when you plan your home-garden plantings fruitfully.

Property borders often pose a problem. Wandering cats and dogs may come to dig and mess your land. Neighbors may not share your appreciation for lovely outdoor living rooms, well planted and tended.

Blackberry hedges along your property line have several values. They conceal unsightly scenes, they thwart wandering animals, and they provide you with luscious berries that are seldom available in any stores today. Raspberry bushes too can form fairly dense rows to provide both privacy and tasty eating every year.

Grapevines that climb and cling to fences have large leaves. These effectively hide areas that you prefer not to see. And as they do, their vines will be producing clusters of red or green or purple, plump, juicy, succulent grapes. With grapes, you have more options. They can be trained to climb up trellises and arbors. With little effort you can build a shaded outdoor sitting arbor. Grapes, with a bit of help in training, will climb the posts, then eventually flow across the top to provide you with welcome shade each summer. The best part comes as your vines begin to bear. Dessert for summer picnics is just a short reach away or hanging beautiful, plump, and ripe overhead.

Perhaps you have your own unsightly corner that has been a planting puzzle. Consider that area for raspberries or blackberries. They can even be allowed to wander in naturalized thickets. Sufficient pruning keeps them in bounds, yet you'll have put that otherwise wasted area to productive use.

Interplanting is important too. When you have lots of room to spare, you can plant where and when you wish. But when space is precious, plan to make every square foot as productive as possible. Berry bushes along a property line are fine. Add flowers in front of them for extra color in summer. If you have stockade or other type of board fences, make them come alive with currants, blueberries, and gooseberries mixed with flowering shrubs in front of them.

Perhaps your driveway needs some added appeal. Most are rather barren stretches of pavement. How about strawberry beds along the edges? Add some dwarf fruit trees. Once you begin to think in multipurpose terms, you'll be surprised how far your creative imagination will lead you.

Evergreens are quite popular around many homes. Blueberries love the acid soil that is enjoyed by evergreens, including azaleas and rhododendrons. Why not plant some blueberries between your evergreens? They'll thrive in the same environmental conditions and give you a bonus with their true-blue fruit.

In parts of Europe, espalier culture of fruit trees and bushes has been raised to an art form. Many veteran gardeners have discovered the fun of training fruit plants to distinctive and unusual shapes. You can too. When you know the secrets, this unique cultural technique can provide eye-catching decorative effects to astound your friends and neighbors. An entire chapter about this espalier growing method is included in this book to guide your pruning shears and saws in precisely the right directions.

Good fences make good neighbors, to paraphrase Robert Frost. What better fences for making friends with neighbors than those living, fruitful ones that both your families can tend and enjoy together? Sharing the cost of the plants and the fun of planting and caring for them and joining in the picking pleasure can bring neighbors together in closer harmony.

The German writer Goethe once observed, "If every man swept in front of his own house, the whole world would be clean."

A famous American nurseryman added to that thought not long ago. He's Paul Stark, Jr., of Stark Brothers in Louisiana, Missouri, our nation's largest and oldest fruit nursery.

"Few of us realize how important an abundance of trees is to man and his environment," Mr. Stark noted. "Right now, we are burning more and more fuel and upsetting more and more the healthful balance of oxygen and carbon dioxide in our atmosphere. For that reason alone, we should have more and more trees because they are nature's principal factories for converting carbon dioxide into oxygen."

Considering trees for home planting, Mr. Stark said, "Trees have many unrecognized values. They add beauty to our homes. Their many distinctive shapes and forms provide variety for more attractive landscape designs. But trees do even more than most people realize. They release moisture that cools the air and washes it. They keep water and wind from carrying away the topsoil that supports all life. They help to keep pollution from washing into our streams. They reduce the silt that clogs brooks and rivers."

Trees have another practical value as well. Real-estate agents will attest to the fact that lovely shrubs and beautiful trees add to the dollar resale value of a home. Time after time it has been proved that the well-planted home with the same square footage, number of bedrooms, and facilities outsells the one with few trees and shrubs.

Trees should be one of your first investments as you begin to plantscape your home. Once they're well planted, they'll be growing and setting permanent, deep rootholds to reward you for many years. As they grow, you can go about other parts of your permanent as well as annual gardening activities, from arranging beds and borders to planting bulbs, flowers, and vegetables.

As you plan, remember these other benefits of trees. They buffer

Thanks to research and new grafting processes, we're able to enjoy wide new selections of dwarf-size fruit trees that bear full-size fruit.

the wind, especially in winter. That saves considerable expense. They also are nature's best noise barriers. Proper planting, according to the U.S. Department of Agriculture and city noise-pollution experts in various urban centers, can reduce noise pollution by as much as 65 percent. That's another reason highway departments are paying more attention to trees and shrubs to muffle roadway noise.

When we talk about planting a tree today, Mr. Stark observes, we're not necessarily referring to a giant oak or a large plot of land. Today, no matter what size plot you have, you can grow a tree. That's abundantly true when you consider the advanced stage of the plant breeder's art and the work done by pomologists with fruit trees and arborists with shade and nut trees.

With research and new grafting processes, we're able to enjoy wide new selections of dwarf-size fruit trees. They stand only 6 to 10 feet tall, yet they bear full-size apples, peaches, pears, cherries, or other fruit. Almost any home has room for one of these marvels of nature made possible by modern plant breeders. You can enjoy several different types or several varieties of one type in the space one large tree would need. The smallest ones, such as dwarf peaches, even can succeed in large patio planters, tubs, or barrels. For that matter, strawberries in window boxes, tubs, and planters are interesting as a strawberry barrel can brighten your home and delight your palate.

As you walk around your garden and home grounds, think fruitfully. Look at the sun that each area receives. Examine the soil to determine how you may improve it. Consider where some berry bushes, a peach, or an apple or even plum tree might prosper to reward you in its tasteful ways.

Specimens of fruit trees can be an accent on your lawn. You can set a miniature orchard along a drive or bordering your vegetables. If room is really scarce, pick your pleasure from any of the fruit trees that may fit a special spot.

Multipurpose landscaping with fruit trees and berry bushes is the most tastefully rewarding way to enjoy the fruits of your garden efforts.

2.
Good Earth Basics

Soil is alive. Even poor soils have their share of tiny organisms, helpful bacteria, and minute creatures at work underground. Every cubic foot of soil, depending on its fertility, can have tens of thousands to millions of beneficial organisms. They have vital functions to perform.

Some devour organic matter, helping to decompose and break down this material to improve the structure of your garden soil. Others work on soil itself, in cooperation with air and water, to break down minerals and other elements. Creatures like earthworms burrow through soil. As they digest organic material, they leave behind castings. This digested material has highly valuable nutrient value for plant roots.

As organic matter is incorporated into the soil, it improves what is known as the tilth. Soil becomes more crumbly or, as scientists say, friable. Air and water and plant roots can move through soil better when it has good condition.

Some gardens already have rich, deep, fertile topsoil. If yours has, consider yourself especially lucky. Many homes and developments have been built on former farmland. More often today, homes have been built on just a light covering of topsoil replaced around buildings after backfilling foundations during construction. Sometimes topsoil has been removed or turned under and less desirable subsoil is on the top level.

Generally you'll find variations of soil, including naturally sandy or heavier clay soils, which are common to certain areas of our country. Don't fret. Whatever you have can be improved. Nature has been at work for centuries building and improving soils.

When you know and appreciate the good earth basics and know how to make soil come alive, you'll be well on your way to growing more productive, rewarding crops from your land, whatever its original condition. There are many simple steps you can take to improve certain conditions rapidly. Other steps will improve soil more slowly, year by year, until it becomes fertile and capable of producing better results than you thought possible. This chapter focuses on those good earth basics in a variety of ways. Since all growing things need soil, your understanding of the good earth and how to improve it can help toward the most fruitful harvests in your neighborhood.

Compost is one of the most useful materials you can make right in your own backyard to improve even good soils. Some people believe compost is something used only by organic gardeners in their zeal to grow plants without artificial aids. That's just not so. Many of the so-called organic gardening techniques are well rooted in solid practicality. Long before chemical fertilizers were widely available to boost agricultural productivity, farmers relied on manure and other organic matter to improve their land. Back on my family farm in New Jersey, we also had compost piles working. I still rely on them. They provide that extra boost for lots of crops, from tomatoes to squash, and help get new fruit trees off to a favorable start.

You can make compost easily and without cost. All you need is a small spot to pile organic materials while they decompose into humus. Or you can speed up the process.

LAYERING COMPOST

There are two basic types of compost making. One is simple layering. This is easier to do, but it will take longer for decomposition to work. In the layering method, you can pick any convenient spot to pile organic matter. A spot out of sight is best, since a compost pile certainly isn't the most attractive part of a garden landscape.

All that is necessary is to pile old grass clippings, fallen leaves, and other organic materials from the garden—pruned softwood branches, debris from weeds, thinned plants—into a heap. You should add about 4 to 6 inches of this type of material, then a layer of soil about an inch thick. You can spread a pound of lime on the pile and a few inches of manure if it is available. If you don't have access to manure, you can spread a few cups of balanced garden fertilizer on the pile. Then add more layers of raked leaves, more clippings from the lawn, and organic debris from the kitchen table, such as lettuce or cabbage leaves. Don't use animal fats or bones, because they will encourage pets or other animals to dig into the pile. Avoid using any diseased leaves. Weeds can be added, since the heat generated by the decaying vegetation will effectively sterilize most weed seeds.

In this layering process, the anaerobic bacteria will cause the decomposition. They work without the presence of air, but they do their work more slowly than do aerobic bacteria. Always leave a depression in the center of the compost pile. If you don't have rain regularly, plan to keep the pile moist by periodic sprinklings with a hose, especially during dry periods. This moisture helps the rotting process.

Layering can be done right around your fruit trees or along your berry bushes too. It takes time but, in the rows around bushes or under trees, this practice can produce several benefits. Mulch layering —the simple and convenient process of spreading leaves, straw, grass clippings, and similar materials around your berry bushes and fruit trees—serves to smother weeds, retain soil moisture, and keep soil cool in hot weather. In addition, the mulch gradually releases small quantities of nutrients into the soil as it slowly decays.

Fact is, many gardeners have a problem these days disposing of leaves in the fall and grass clippings and plant prunings during the growing season. Many towns now ban the burning of leaves and brush. It takes time and money to get lawn trash bags and haul these materials away. Often you can ask neighbors for their extra leaves and clippings for your mulching and composting plans. If the neighbors don't want these materials, you both win.

INDORE METHOD

The other compost method, known as the Indore method, gives you considerably faster results. It is based on the program developed in England many, many years ago by Sir Albert Howard, the father of modern organic gardening. You don't have to be an organic gardener to follow it. Rather, it pays to use the best of all different gardening techniques wherever they originate around the good planet Earth.

The Indore method has been revised in several ways over the years. All the variations involve turning the compost material periodically so that the faster-acting aerobic bacteria can decompose the material quickly. They work best in the presence of air. You can do the turning

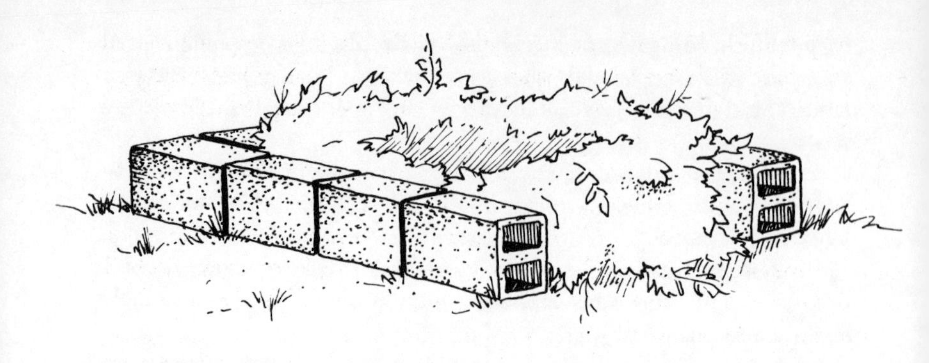

One of the simplest ways to build a backyard compost pile is to start by arranging tiles or cinder blocks as a three-sided bin.

Another easy way to start a compost pile is to arrange a section of snow fence or wire fencing in a cylindrical form.

by pitchfork, by spade, or mechanically. By placing compost material into bins with perforated sides or providing wire frames through which air can move more easily, you will encourage this speedier decomposition.

In the accompanying drawings, you see two of the easiest ways to build a simple backyard compost pile. In the first, tiles or cinder blocks are arranged to provide a three-sided bin. In this you can pile all types of materials. Then, once or twice a week, or more often if you have time, turn this material. Keep it moist so the anaerobic bacteria also can work.

Adding manure from cows, sheep, horses, or poultry incorporates nitrogen and small amounts of other nutrients. That's the preferred method of organic gardeners. You can add several cupfulls of 5-10-5 or 10-10-10 commercial fertilizer to help speed up the decay and add those extra nutrients in the finished compost.

By following this regular aerating method, you'll be able to get finished compost within two to three weeks if moisture is adequate and the weather is warm. All types of compost activity slow down in dry or cold weather.

The second easy way to establish a compost pile is to use a section of snow fence or wire fencing and set up a sizable round "bin" in which you can accumulate organic material.

HOLE COMPOSTING

From parts of Europe where topsoil is sparse and of poor quality, veteran fruit growers brought another soil-improvement tradition to America. They knew how to improve soil for fruit trees, grapevines, and other berry-fruit plants simply in holes in the ground. You can take a tip from this method. It's really making compost in a hole.

Consider the spots where you want to plant your trees or bushes. Do they have sufficient good sun, but soil that isn't to promising? Well, dig into the ground. Make a hole at least twice the size of the normal spread of the roots of your intended plant. It is best to do this in the fall before planting.

After you remove the soil, begin placing leaves, manure, and peat moss into the hole. Add grass clippings and kitchen refuse. Add fertilizer, about a cup or two, depending on the size of the hole. Add some topsoil, an inch or so to press down the lighter leaves and grass.

Use a spading fork or spade to turn the material periodically. Keep it moist, of course.

If the soil you have removed is rocky but has a fair amount of organic matter and seems reasonably fertile, screen out or pick out rocks and other debris and save the soil. You can mix it into the hole with the compost material.

If you make compost in a pit, you won't have unsightly piles around your yard. Aeration isn't perfect, but you'll be able to see how material rots down into the hole. This porous, organically enriched base is excellent to encourage root penetration and provide better drainage and moisture-holding capacity when you do the actual planting the following year.

One area of our garden in New Jersey some years ago had exceedingly poor soil and even worse subsoil. Much of it seemed typical Jersey red-shale soil. To improve the entire area quickly would have involved extensive work. So I tested this compost-hole system. It provided a wonderfully enriched base for a row of blackberry bushes. The next year, I made more holes. For several years I continued to improve the area, several compost holes at a time. By the end of four years, that entire berry area was thriving. During each year, I also mulched with as much straw, old hay, or leaves from fall raking as possible, up to 6 inches along rows and around bushes.

TRY GREEN MANURE

If you have a chance to plan ahead for fruit trees and berry bushes, do so. Often a "green manure" crop of winter rye, clover, or similar plants can set deep roots to open the soil for a year before you plant that berry patch, grape arbor, or grouping of fruit trees. Legumes are best. Clovers have a unique benefit. They can fix nitrogen from the air on their root nodules. When you dig under or till under these legume green manures, they release natural nitrogen freely to benefit your fruit crops in that spot.

Interplanting also has its values. The first year, berries are usually slow to start. It may take several years for them to fill in an area, either along the row or as a thicker hedge for a property border. The space between rows shouldn't be wasted. You have a wide choice of other crops that can provide bountiful yields for one or two years between rows or even in front of them until bushes fill in.

We have favored cucumbers and squash. The vines creep along the ground, shading it and providing an interesting ground cover. We also enjoy the squash and cukes from these interplanted crops. Lettuce, broccoli, and even tomatoes are suitable for interplanting. However, when you interplant, you must make allowances for the extra crops. They too need their own nutrients and moisture to reach prime maturity and begin setting crops.

Soybeans and snap beans or limas work well. They take little space and, being legumes, they also can help improve the soil, adding their contribution of nitrogen from roots to improve soil for future use by trees and berry bushes.

If you decided to set aside an area and improve it for berry bushes, here are some basic pointers. When you can, deep-till the sod 8 to 18 inches deep. If there is no sod, you can start by spreading leaves, grass clippings, or similar material on the surface. When manure is available, spread a 1- or 2-inch layer on the ground. Dig or till it under, working it into the soil.

After the seedbed is raked smooth or tilled evenly, plant your winter cover crop of green manure. Rye is fine. It roots well, and its roots open the soil for spring tilling just before you set your berry patch or fruit-tree hedge.

If your soil is respectably fertile, and you don't choose to plant green manure, you have your choice of preparing soil in the fall or just before spring planting. You'll note in the next chapter (Planting Guidelines) and in other chapters that some fruits do best when planted in spring; others prefer fall planting. The tender fruit crops, peaches and their relatives in particular, do best when spring-planted.

If you dig or till the selected site in the fall but will wait until spring to plant, spread some fertilizer on the surface in the fall. A pound of 10-10-10 or 10-6-4 per 20 square feet will work into the soil over winter as moisture spreads through the soil. Most of it will be there when needed at spring planting time, although some will be used in helping decompose organic material in the soil. A minor amount is lost, of course, through leaching, especially in sandier soils.

Perhaps a word is in order about improving two other types of soil problems, whether you plan fall or spring planting, fall soil preparation, or whatever system your time and planting plans dictate.

Sandy soils tend to dry out. Many areas, especially coastal piedmont regions of the eastern states, have sandy soils. These don't hold moisture too well. Neither do the sandy soils of the Southwest. In dry summers, shallow-rooted trees and bushes can suffer badly in sandy soil. Fortunately, there are several easy ways to improve these soils quickly. Organic matter is the key, but you don't necessarily need compost. Peat moss is available from garden centers everywhere. It is one of the most versatile products to improve soils, especially sandy types. The best buys are the 6-cubic-foot bales.

Mix a bushel of peat moss to every estimated 2 bushels of sandy soil. Work it in well. Then, after planting, apply about 2 inches of peat as mulch.

In the soil, peat improves moisture-holding capacity of sandy soils. Applied to the surface, it retards evaporation and smothers weeds so they don't pull moisture from the soil to rob your plants of that vital element.

In soggier soils, those composed of large amounts of silt and clay particles, plants have another problem. Most plants, including trees and berry bushes, can't stand wet feet. That's understandable when you realize that plant roots require air to breathe. Without air movement and adequate transfer of nutrients as well as ability of tiny feeding roots to penetrate soils, plants won't prosper. Clay soils in dry weather often form hardpans—tight, hard layers, often below the surface, that thwart root penetration. Sometimes you can see this problem right at the surface as soil cracks in dry, hot weather. Even with these soils, which may occur just on a portion of your property, usually a lower-lying area, you can change the texture and quality considerably. Here again, organic matter and peat especially play an important part.

Peat moss can be incorporated into clay-type soils. But don't do it when they are wet. Wait until they are somewhat dried, so digging or tilling won't compound the problem by forming cloddy clumps. Mixing sand and peat into clay soils is an excellent practice. Use a shovelful of sand with four to five similar amounts of peat moss for 2 to 4 square feet if soil is heavy clay. Spread it on the surface and dig or till it in. After the first rain, check to determine how much you have improved the drainage. You may need to do this several years in a row to thoroughly improve the particular area as your plants are growing.

Composted humus, sand on the surface as mulch, and manure also can be spread on clay soils. Each year you'll see improvement as you incorporate organic matter into the ground. Soggy soils are more difficult to improve than sandy ones, but you can succeed, quite remarkably.

Nature will work her wonders with your help. As you improve soils, you'll notice that they physically come alive. To prove this fact to yourself, dig a shovelful of soil near a compost pile or an area that has been well mulched. You'll find little creatures, especially earthworms, at work. It has been said, and rightly so, that when you find lots of earthworms in soil it is really healthy.

Another fact about soil is abundantly true: in its natural state, it has a profile. That won't hold true if you are forced to garden on backfill around a home or building, but soil in natural areas does form in a systematic way. You can look for the profile in a desired planting area by digging down with a spade. If the land hasn't been touched much, you'll find a clearly visible profile with several horizons.

The upper level is the topsoil. It is a combination of the broken-down minerals and bits of gravel from the subsoil with the organic matter that has been dropped by living plants and decayed into it. Topsoil is usually darker than the next level below.

Below the topsoil is the subsoil. It is usually more gravelly and lighter in color. Below this is the parent material, which may be anything from rocky soil to shales or bedrock, depending on where you live. The profile itself can be shallow, with just a bit of topsoil. Or the profile can be deep, with rich layers of topsoil extending several feet down as in the Great Plains areas. Soil is formed slowly over tens of thousands of years. If you have good, deep soil, be thankful. But don't despair if your home isn't blessed with good topsoil. All soil can be improved. Even fertilizer added to sandy soils will provide necessary nutrients to keep plants growing.

BE CAUTIOUS IN BUYING TOPSOIL

Perhaps a warning is in order. It may seem logical to improve a garden area for berries, fruit trees, and even vegetable and flower gardens by purchasing topsoil. It pays to have a soil test made on topsoil that you may be thinking of buying, to be sure it has a reasonable fertility and isn't merely fill.

Several other facts about your soil should be considered as you begin your fruitful landscape plantings. First, learn about the types of soils.

Texture refers to the size of the majority of the particles making up the soil. It ranges from those tiny, almost microscopic clay particles to small stones or gravel.

Clay soils can be stony clay, gravelly or sandy clay, silty or just plain muddy clay.

Loamy soils may be coarse sandy loams, medium, sandy, fine, silty, or clay loams with an abundance of clay particles in them.

Sandy soils range from gravelly to coarse to medium, fine, and loamy. These are the terms that your county agent may use as he helps you evaluate your sites and planting areas.

Structure of soil is determined by the way in which individual particles are grouped. A good structure lets plant roots, air, and water move freely through it. Loamy and clay soils may have a crumbly structure. Sandy soils have little granulation. Clay soils will compact readily.

The easiest test to evaluate soil structure is this: pick up a handful at planting time and make a fist. If it crumbles easily after you have squeezed it in your fist, the soil is probably the desirable sandy loam that will perform quite well. The closer you can improve the structure to a granular feel with clusters of soil that easily shake apart, the better. Organic matter, from peat to compost and humus, will keep you on the track in your soil-improvement plan.

One final point should be understood as you select your planting sites and go about whichever soil-improvement program you want to make your land more productive. Three kinds of water are found in any type of soil.

The first is *gravitational water*. In sandy soils water often drains out too quickly, leaving plants to wilt or perform poorly. In clay soils, gravitational water lingers longer to create soggy spots, with soil pores clogged with water. That will rot roots.

Hygroscopic soil water is that which is chemically bound with soil materials. It is basically unavailable to plants, so it's of no great concern to you.

The most important soil moisture is known as *capillary water*. It is free to leave the soil and enter growing roots. As it does, it carries plant foods into the plants, up through stems or trunks into branches, twigs, and leaves. This water is most available when soil texture and structure are crumbly and loamy with ample organic matter incorporated into the soil.

Entire books have been written about soil. This chapter is meant as a primer. The good earth may be waiting your planting of these fruitful trees and bushes now. If not, heed the hints in this chapter to improve it in the variety of ways outlined here. Soil is truly alive and can be made to come more alive and healthier.

3.
Planting Guidelines

One of the important things I have learned from appearing on TV, giving my good-growing tips as America's Green Thumb Gardener to the millions of people who watch each month, is the need for brevity. In a book, even the most careful authors tend to ramble on, lost in their natural exuberance and enthusiasm for their subject. In this book, I have tried to be as clear, concise, and accurate as possible, despite my natural enthusiasm for this topic. Having helped farm 90 acres of apples, pears, and peaches on our family's farm years ago, I also planted, pruned trees, and harvested a few thousand bushels of fruit.

Home gardening is much more fun. You can take time to do each job better. The rewards certainly are sweeter. However, brevity was the point. Here's a simple, down-to-earth checklist for selecting, planting, and tending fruit trees and berry bushes. It's brief and to the point. Other details are provided in chapters on the different trees and bushes. But for easy reference, this checklist will be helpful. I owe a debt to the fine people at the New York State Fruit Testing Cooperative for making me rethink in briefer terms at times. My thanks to them for this suggestion, which is based on their exceptional knowledge of fruit planting.

READY-REFERENCE PLANTING TIPS AND FRUIT PLANT CULTURE GUIDE

1. Everything begins in the soil. Most fruit trees thrive best in deep, well-drained, and friable (crumbly) soils. Pears and plums will do well in heavy soils. So will currants and gooseberries. But peaches, cherries, and strawberries prefer a lighter soil, a sandy loam.

2. Age of plants is important. The youngest available usually transplant best and are the least expensive. One-year peaches and sweet cherries are most satisfactory. You should order one- or two-year-old trees of apples, pears, and plums. Don't waste money on older trees.

3. Planting time varies for types. Fall is best for sweet and sour cherries, except in cold areas. Apricots, nectarines, peaches, and plums

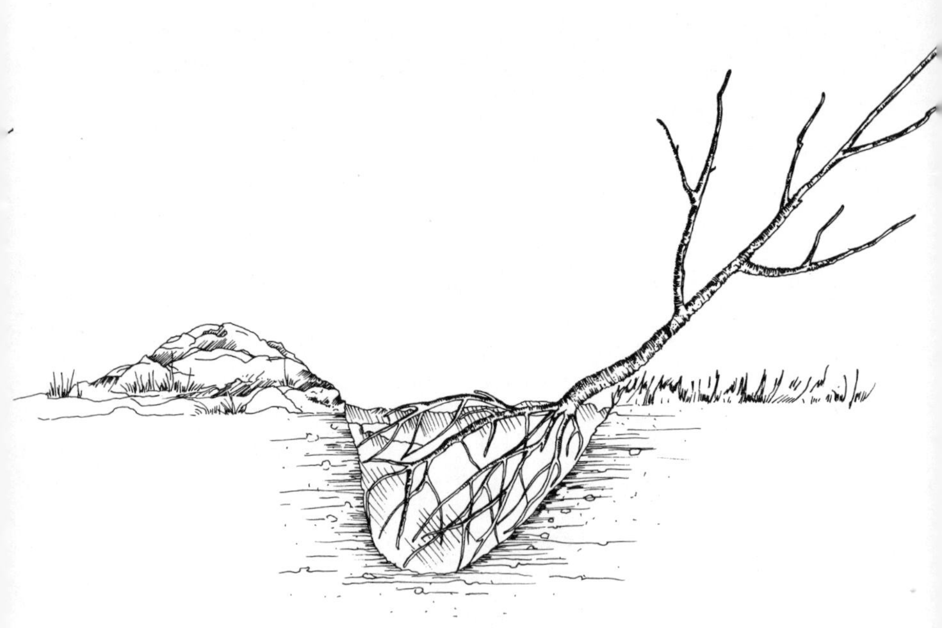

If you can't plant trees or shrubs immediately, you can heel them in for a week or two with their roots beneath moist soil in a shady spot.

should always be planted in spring. Apples, pears, grapes, and other small fruits can be planted in spring or fall, but spring is recommended by most experts. If you select spring, do it early so plants get a good start and have extra growing time during good weather.

4. Selecting the right site is important. Plant where you like for eye appeal, convenience, and other personal considerations, provided the soil is good. If not, improve it. But avoid frost pockets, low areas where frost settles, and be sure your site is blessed with ample sun.

5. Pollination is necessary, naturally. Without it, fruit set is reduced or doesn't occur. It is good insurance to have an least two compatible varieties that pollinate each other in the same vicinity. Peaches and sour cherries, however, are self-fruitful. They'll pollinate themselves properly. Pears usually need a nearby pollinator to set full crops.

6. Variety selection is up to you. However, look over catalogs carefully. Some old-time and new varieties offer special advantages, from great taste to disease resistance. Some are better for all purposes than others, which may be best for freezing or fresh use but not good for both.

7. Consider multiple values. That's the purpose of this entire book, to focus your home landscaping plans toward more fruitful rewards.

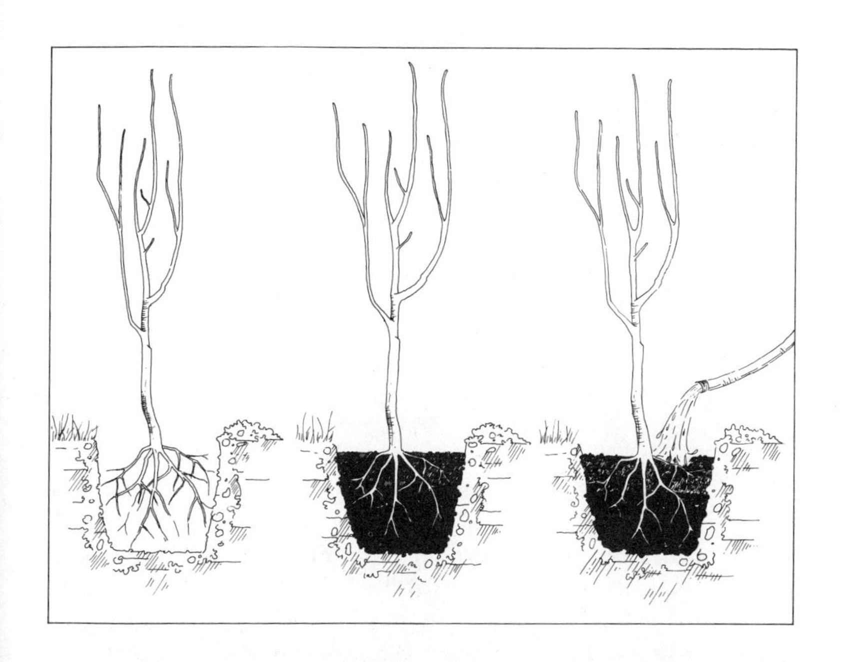

Any fruit tree gets off to a better start when you plant it in a hole of the proper size, add an improved-soil mix, and keep the soil watered.

Most fruit trees can be used as ornamentals as they produce their bonuses for your table or freezer or for other purposes.

Blackberries, raspberries, grapes, blueberries, and elderberries all make fine hedges. Attractive hedges you can eat are doubly valuable. Currants, gooseberries, and blueberries are well suited for borders or even specimens (singly or in groups) or interplanted with other shrubs and trees. You can enjoy fruit trees as specimens for blooming spring color, summer or fall fruit, and foliage displays. Ornamental crab apples fit well in many spots, especially the double flowering kinds. Nut trees, too, can provide shade, pleasing perspectives, and their tasty crops of nuts to nibble each year.

8. Bear with your trees and shrubs. They'll bear well for you when you plant and tend them properly. Learn what to expect and when. Small fruits begin bearing within two to three years, depending on when you plant them and on the growing conditions those first years. Grapes and peaches take three to four years before they begin rewarding you with fruit. Plums and cherries take five before they reach productive maturity. Apples and pears may require five to eight years, depending on the variety. Dwarf fruit trees bear earlier, usually several years earlier than standard-rootstock trees. Remember, to stretch your harvest season, select early-, mid-, and late-bearing varieties. Handy charts on bearing age, productive capacity, and other helpful information are included for your convenience and ready reference in reading fruit nursery mail-order catalogs to help you select the best early-, mid-, and late-bearing varieties.

9. Planting isn't a big problem. It's easy, but don't rush through it. Open your plants when they arrive. If you can't plant immediately, heel them in so roots are tucked safely beneath moist soil in a shady spot. If roots have dried, soak them in a basket of compost and water, a cup per gallon as a slurry. Always keep roots from drying in the sun while planting.

Be sure to dig the hole large enough to give roots space to spread naturally. Restricted planting slows growth and may delay your joyous harvest by a year or more. Poor planting can cost you a tree or shrub. Before planting, prune damaged parts and broken root tips. Pack soil firmly. Water well. Standard trees should be set slightly deeper than they grew in nurseries they came from. Dwarf trees should have graft unions at least an inch or two *above* the soil. Bush fruits usually should be the same depth as or slightly deeper than they were in the nursery. Appropriate chapters give details on each fruit in this book.

10. Pruning is perhaps, after proper planting, the single most important factor in producing an abundance of fruit. It stimulates the growth of shoots and branches and canes, resulting in new fruiting wood that bears your expected bounty. Read carefully the details for each type of tree and shrub. The better you prune, the more success you'll have with fruit plants. Pruning is the most overlooked but vital stage in fruit production. Simply cut tops of new, young trees back about one-half their length. Leave two or three well-spaced branches and—except with peaches and their relatives—a leader. Remove narrow crotches. Make pruning cuts to an outside bud so new branches grow out, not into the center of trees.

11. Mulch is a boon for good gardening. Collect and compost old

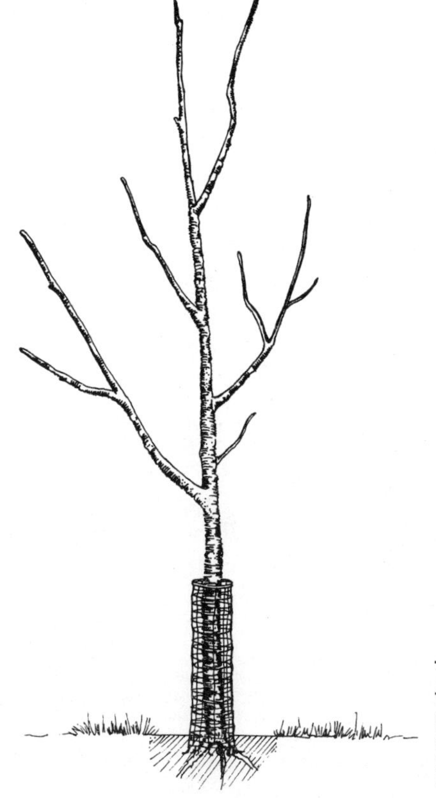

The way to prevent rodents from girdling the bark at the base of young fruit trees is to install wire mesh in this fashion, not touching the bark.

leaves, grass clippings, straw, and similar organic materials. Use them as mulch, or, if composted, use compost as mulch. Mulch saves moisture, smothers weeds, keeps fruit bushes and trees tidier, and adds minute amounts of nutrients to benefit their growth.

12. Feed your plants if you expect them to feed you in time. Nitrogen is the most important element, but balance of all essential nutrients on berries is important too. Trees can do well on good soil by themselves, but for the most abundant crops, year after year, you must put back into the soil what your growing crops take out.

13. Pests play a part in every effort to grow fruit trees and shrubs, bushes, or plants. Rodents can girdle young trees. Wire mesh around saplings keeps them away. So do poisons, but use only those that won't harm pets or children. Insects and diseases can thwart your best efforts. Resistance to disease has been built into improved varieties by plant breeders. New improved pesticides are safer to use and can help you conquer even the most stubborn problems. Play it safe and always double-check the manufacturer's directions. Be careful while mixing and applying. Keep chemicals away from youngsters always.

14. Know how to tell when fruit is ripe and ready. Details are included in the chapters on individual fruits.

4. Apples Are Appealing

Apples are as American as apple pie, thanks to good old Johnny Appleseed. He did his job well, scattering seeds of this favorite fruit far and wide across our growing nation.

There was a time when just about every home had an apple tree or two, it seems. They were extolled in song and story. Today, largely because of surburban sprawl, thousands of acres of apple orchards are gone. The bulk of our apple crop is produced in commercial orchards, which concentrate on high-yielding, easy-to-harvest, pack-and-ship varieties. Gone from supermarket shelves are many of the tastiest, juiciest, best pie-making and baking varieties.

Fortunately, with the renewed interest in home vegetable gardening has come rekindled interest in home fruit growing. Luckily, many of the best among the old-time favorite varieties were still retained by some nurseries. They are available today to provide far more flavorful apples than you can buy in stores.

You can grow Northern Spy and Winesap or even such exotic delights as Winter Banana right in your own backyard. Fact is, some of the tastiest varieties can be grown only in home gardens. That may sound strange. Perhaps a bit of agricultural background is in order to explain why.

Millions of home gardeners have returned to growing vegetables to fight inflation in recent years. When they did, they realized that home-grown vegetables are juicier, more flavorful, more nutritious. Many gardeners attributed this superiority to the simple fact that their home-grown produce was fresher. What most people don't fully understand is why commercially grown vegetables and fruits may not be as tasty as home-grown vegetables and fruit.

As our nation's agriculture expanded to feed our growing population, vegetable farmers focused their attention on certain varieties that could withstand long-distance shipping well without bruising or spoiling. They sought varieties that would retain color and appearance. The same factors are true for commercial fruit growing.

Many of the tastiest varieties didn't have these vitally important characteristics for efficient, large-scale commercial production. Consequently, more flavorful older varieties were eliminated as commercial orchards were replanted with those that provide the necessary attributes.

Luckily for all of us, enough demand remained among veteran gardeners for nurseries to retain sufficient rootstocks of old-time apple varieties. Today, we can again enjoy them as part of our individual fruitful landscapes.

Just as important to the rebirth of home fruit growing are the new developments in plant breeding plus advances in safer chemicals to control troublesome diseases and fruit insects. Dwarf and semidwarf rootstocks make it possible to fit several trees into the same growing room that a standard tree requires. This development more than any other single factor has led to the apple tree's return to home gardens.

Apples are again a natural for multiseason beauty just about anywhere you want them. When apple-blossom time arrives, they perfume the spring air. Their delicate pastel pink blooms are harbingers of spring as much as the daffodils blooming beneath them. Their attractive shapes and distinctive growth patterns fit tidily into landscape plans. Come fall, the bright red, golden yellow, or apple green fruit of different varieties add colorful beauty to autumn scenes.

As you plan your apple orchard or specimen-tree plantings, consider the unique advantages of dwarf and semidwarf trees. They may be small in stature, but they do bear full-size fruit in sufficient quantity to satisfy most families.

Dwarf apple trees are the result of grafting the desired flavorful varieties on special rootstocks developed originally at the East Malling Fruit Research Station in East Malling, England. Over the years, plant breeders have discovered and cultivated different types of dwarfing rootstocks that permit different degrees of dwarfing. As you read nursery catalogs, you'll find they indicate the type of rootstock available and the size of the fully mature tree that will grow on each type. These rootstocks are identified by Roman numerals.

Malling IX has been widely used in all parts of our country. It is also used commercially where orchardists wish to contain the size of their trees for ease of spraying and harvesting. This is the best of the fully dwarfing rootstocks. It keeps the apple tree no taller than 8 or 9

Today's apple trees on dwarf and semi-dwarf rootstocks make it possible to have several trees in the same space that one conventional tree would occupy.

feet over a twenty-year period. That's handy. You can easily prune and pick apples from these trees while standing on the ground. Trees on Malling IX stock are useful for very small plots. You can use them on front lawns or as compact specimens at the corners of your house. Trees on Malling VIII also are dwarf and perform about as well as the IX ones. Some nurseries offer apples on other types of dwarfing rootstock that work as well.

Since these trees take little space, you can plant a hedgerow of them along a driveway or property border. They neatly screen out unwanted views, yet remain easy to trim and tend without fear of overgrowing. True dwarf fruit trees have a tendency to overbear in their urge to provide you with an abundance of fruit. That tendency usually derives from the bearing variety. Another complaint about dwarf trees is that they are somewhat shallow-rooted. With a heavy fruit crop in season, or when laden with ice during winters, they may tip and uproot. These problems can be easily corrected by thinning the fruit and bracing or staking the trees.

You can enjoy these truly dwarf apple trees in groups in a corner of your yard. Combination plantings let you enjoy several varieties with their slightly different blooms in spring and contrasting fruit colors in fall. When you realize that they can be planted as close as 6 to 8 feet apart in rows or groups, compared to the 35- to 40-foot spacing required for one standard apple tree, you'll be able to visualize their multiple uses around your garden and grounds.

Malling II is another popular semidwarfing rootstock used by many nurseries to produce smaller apple trees. It is hardy and encourages early and prolific bearing. This rootstock will produce mature trees about the size of a cherry tree. *Malling VII* is similar and has the advantage of somewhat greater drought resistance. It proves more vigorous and hardy than II.

Malling I, II, III, IV, V, VI, and *VII* are all dwarfing stocks. However, VIII and IX are considered the best for producing truly dwarf trees. Some of the lower numbers have shown a tendency to develop suckers or make a poor union with some apple varieties. The Clark dwarfing stock being developed in Iowa has promise. It is being used as an intermediate stem between root and top variety. Some nurseries offer this in addition to the Malling stock.

Another method of producing semidwarf trees is grafting. For example, a hardy standard rootstock may be used on which a *Malling VIII* interstock is grafted in the nursery, with the desired variety grafted on top of the interstem. According to plant breeders, the degree of dwarfing is apparently increased with the length of the interstem used. Other nurseries offer dwarf trees on other, newer dwarfing rootstocks too.

In general, the interstem method of creating smaller trees usually produces trees about one third to one half the size of standard fruit trees. Research is continuing on these procedures to produce more convenient sizes of apple trees.

As you shop for dwarf fruit trees, check with your local nurseryman or read the mail-order fruit catalogs carefully. That way you'll be certain to get the stock that will produce a tree of the size and shape you want so it fits properly into its desired place in your landscape.

Dwarf trees have other advantages. They begin to bear earlier than standard-size trees, usually two or more years earlier. You can expect about a bushel of apples from a dwarf tree about the size of a mature standard peach tree. Early bearing is appealing, since you can enjoy apples sooner. Dwarf trees have a life span of approximately twenty to twenty-five years before they begin to decline in vigor.

Dwarf trees also let you plant several varieties of fruit with differ-

Dwarf apple trees start bearing fruit sooner than standard-size trees do, producing about a bushel of apples per tree of the size shown here.

ent seasons. Some will bear early, others in midseason, and others in late season. That stretches your apple-eating harvest over more months. Pruning and care are easier with smaller trees. You eliminate those tottering tall ladders when pruning and picking your crop.

If you do have lots of room, of course, you can plant standard trees. Some of these, and dwarf or semidwarf types too, have been grafted with several varieties. As trees begin to bear, you can pick Red Delicious, McIntosh, Yellow Delicious, and others right off the same tree. The grafts of the different varieties by expert nurserymen have created these 5-in-1 apple-growing wonders. Stark Nurseries offers them in their catalog. In general, it is better to grow several desired variety trees, since care of a 5-in-1 is a bit tricky, especially pruning.

You will find in Chapter 20 a source of reliable mail-order nurseries (Reputable Suppliers of Fruit Trees, Shrubs, and Plants). They provide a wider selection of varieties than most local nurseries can supply. A discussion of apple varieties can become complicated when you realize that there are more than 2,000 of them. Many began as chance seedlings. However, only about 100 of these are of commercial or home-garden importance today.

Which varieties you select will depend on your own individual taste, of course. For convenience, here are some of the best for eating fresh, for multiple use, and for pies and cooking. Eating homemade applesauce, by the way, can be a surprisingly powerful experience compared to sampling those jars and cans you buy in stores.

Quinte is very early and hardy. The fruit is medium-size, red with yellow streaks. The flesh is soft, aromatic, tender, and great for dessert.

Prima is early midseason. The tree is moderately vigorous and spreading. The fruit is medium-large with 60 percent dark red color on bright yellow background. The flesh is fine-grained, firm, crisp, and excellent for fresh or cooking use. This variety tends to resist apple scab disease.

McIntosh is popular for home gardens. Trees are hardy and vigorous. The fruit is medium-size, bright red, blushed with carmine strokes. The flesh is firm, crisp, and very juicy. Variations of the old McIntosh have been developed. Rogers and Red McIntosh are two good ones.

Macoun is a midseason ripening variety. Trees are upright, hardy, and bear medium-size, dark red-striped fruit. The flesh is semifirm, crisp, and white, excellent for fresh use.

Spartan also is midseason with upright vigorous growth. The fruit is solid dark red, firm, crisp, juicy, and white. This apple is more resistant to preharvest drop than McIntosh.

Cortland ripens in midseason with semifirm, crisp, juicy fruit. It is noted for white flesh that does not brown readily and thus is good for dessert, cooking, and salads.

Priscilla ripens in midseason. It is conical to round in shape, glossy, with white to slightly greenish flesh. It, too, is reportedly nearly immune to scab as well as resistant to fire blight and powdery mildew. That makes its care easier.

Idared ripens late on upright, very productive, vigorous trees. The fruit is medium to large, bright red with creamy white, firm flesh, good fresh and for cooking. It retains quality under storage long after harvest.

Mutsu, a Japanese introduction, is very vigorous and productive. It bears large golden yellow fruit late. The flesh is yellow-white, crisp, and good for many uses. It is a good variety to replace Golden Delicious because it is free from russeting and its leaves have resistance to fruit-spray injury. It, too, has long storage potential, but must be grown with other varieties to ensure proper cross-pollination.

Some apples are more susceptible to insect and disease problems than others. Early maturing varieties such as *Lodi, Summer Rambo,* and *Grimes Golden* usually are attacked less by insects and disease than varieties which ripen later. One main reason is that fungus diseases are favored by warm weather and moisture, which usually occur later in the summer. Other varieties have been developed to beat pest problems by inbred genetic factors. Fruit catalogs usually spotlight varieties with such useful characteristics.

If you love apples, you should join the New York State Fruit Testing Cooperative Association in Geneva, New York. It has a long record for developing, testing, and introducing exceptional new tree and bush fruits. Membership is five dollars per year.

As a member, you can buy new test varieties for evaluation in your home garden. You can also get a wider range of apple varieties than you can from most other sources. The association offers stocks of *Burgundy,* a blackish-red apple; *Ozark Gold,* a Golden Delicious type; *Vista Bella,* one of the very earliest red apples; and some of the old-time exotics.

For the apple connoisseur you can find *Winter Banana* or *Rhode Island Greening, Red Seek-No-Further* or *Hubbardston Non Such.* You can also get such great-tasting but strangely named apples as *Chenango Strawberry,* with its long, conical shape; *Esopus Spitzen-*

burg, with a heady aroma and orangy-red color; and *Red Astrachan,* a beautiful early summer apple. The catalog is issued yearly and is a handy guide wherever you shop for apples or any other types of fruit trees and berry bushes. The Stark Brothers and Bountiful Ridge Nurseries also have excellent, well-illustrated, free catalogs.

CONSIDER CRAB APPLES TOO

Don't overlook the beauty of flowering crab apples in your living landscape. Conventional single-flowering crab apples are hardy and provide profusions of bloom. They reward you with those marvelous bright little apples for making jelly that just can't be duplicated by jelly manufacturers. Decorative double-flowering crab apples, on the other hand, offer greater profusions of bloom. They vary in form and size from the shrublike *Sargent* to narrow upright *Pink Spires* and the wide-spreading *Lemoine.* Some are weeping, others vaselike; still others are more irregular and picturesque.

You can get flowering crab apples with white, pink, or red blooms. They may bear single flowers with five petals or semidouble and double with clusters of petals.

The fruits of flowering crab apples may range from pea-size to multipurpose ones 2 inches in diameter. Their color is generally red, purple, or yellow. If you wish to fit crab apples into your planting scheme, select flowering varieties on the basis of their fruit display rather than flower array alone. Flowers are colorful for one week to ten days. Fruits last up to 6 months unless you pick them for jelly making. They also have the advantage of being excellent food for your feathered friends during those long, cold winter months. Smaller-fruited varieties produce less litter if you don't plan to use them for cooking.

Consider the particular crab apple variety's susceptibility to diseases too. *Hopa* is popular and produces large fruit. However, it is most susceptible to scab, a common disease of apples. Scab can be transferred to other apple trees if it becomes established on a susceptible crab apple variety.

Redbud is one of the best all around. The buds are pink, the flowers white, and the pea-size fruits last all winter. *Adams, Profusion* and *Red Splendor* are better than Hopa. They have reddish foliage and reddish-pink flowers. Fruits are more colorful. They are quite resistant to apple scab.

PLAN AHEAD

Before you decide where you want your apple trees, consider the shape of things to come. Apples can be trained on a trellis, especially the dwarf types. This is called espalier culture and is an art by itself. You'll find an entire chapter on this exciting and easy-to-accomplish tree training system in Chapter 16, Space Sculpturing Is Tasteful Too.

Apples like sun and reasonably fertile, well-drained soil. Once you spot the ideal location where the beautiful delicate blooms and sweet perfume will waft across your garden, prepare the soil for planting. It is best, as with all trees that will become permanent parts of your home grounds, to dig in deeply and do this job right. A little extra effort at planting time will assure your tree of a happy home where it can set its roots deeply and well for a long, prosperous life in that location.

Fall is the best time to prepare soil. When trees arrive in the spring from your favorite mail-order nursery or local garden center, you will be prepared to put them in the ground, even if the weather is not yet perfect. That way, you get a jump on planting, before other garden chores press you for time.

In milder areas of the country, if temperature is seldom below zero, trees may be planted in fall, winter, or early spring—anytime the ground is not frozen. Apple-tree roots are more tender than tops. Therefore, fall planting in colder areas can be done more successfully with apples, pears, sour cherries, and European types of plums than with peaches, sweet cherries, and Japanese types of plums.

Fall planting has some advantages over spring planting. Often soil is in better condition. Weather also may be more favorable, as fewer windy days or long wet periods occur in fall than in spring. A fall-planted tree also has time to get a good roothold and be ready for strong spring growth when the first blush of warm weather arrives. That gives you almost a year's growth advantage toward earlier bearing.

IMPROVE THE SITE

Organic matter is valuable for all growing things, whether trees, shrubs, flowers, or vegetables. Plan to mix into the soil whatever organic material you have or can get. Use peat moss or manure, compost or leaf humus. These materials will open heavier soils to let roots, water, and air penetrate better. They will also improve the moisture-holding capacity of sandier soils to avoid drought damage in dry periods.

Manure and compost are usually more readily available in fall than in the spring peak sales season for local suppliers. And prices may be lower. You may also be able to get manure from riding stables or farms in the fall, free for removing it, rather than pay for manure when everyone wants it in spring.

Whichever season you prefer for planting, remember that fruit trees should be set before the middle of April, while they are still dormant, not yet leafing out. Most trees arrive from nurseries bare-rooted. You can buy container-grown or balled or burlapped trees locally, but they usually cost somewhat more.

Young one- or two-year-old plants are the best buy. You can buy older plants, but tests have shown that often they don't begin bearing any sooner since they may be subjected to greater transplanting shock.

PLANTING POINTERS

You can dig the holes for your trees by hand or prepare the ground by rototilling. I consider a rototiller the second most important power garden tool next to a lawnmower. It can dig holes for trees, beds for shrubs, and gardens for flowers and vegetables. It can also be used to cultivate and turn under anything from sod to layers of organic matter for improving the soil. You often can rent them.

If your soil is poor, by all means improve it. Details on this vital practice are included in Chapter 2.

Give your trees enough room to grow naturally. Position full dwarf apple trees 10 to 12 feet apart each way or 6 to 8 feet apart in rows with 15 feet between the rows. That's ample room to let them mature properly and tend them well.

Place semidwarf apple trees about 20 feet apart each way. Standard trees will need 35 to 40 feet between them to reach their full growing potential without crowding. If you plan to create hedgerow effects, you can plant all sizes closer than these suggested distances. Pruning will be somewhat more complicated, but as property borders and living screens to block out unwanted views, fruitful hedgerows are functional as well as attractive.

Before you plant any tree or bush, consider its size at maturity. Keep trees and shrubs far enough away from buildings and property lines to avoid maintenance problems. Remember, too, that fruit trees and bushes may require periodic spraying. Your neighbors may object to spray drift on their property, although this is less a problem than it may seem. Proper spraying can keep drift to a minimum.

Dig each hole large enough to accommodate the entire root system without crowding. It should be deep enough to allow the tree to be planted at the same depth it grew in the nursery. You can usually see that depth indicated by a different color and texture of wood on the trunk. While you dig, separate the topsoil from the subsoil. If the soil is good, spread some of the topsoil in the bottom of the hole. Spread roots gently over it. Then sift more topsoil around the roots. Add improved soil that is combined with manure or compost as indicated in Chapter 2. If the soil is extremely acid, mix 1 to 2 pounds of limestone with the soil in the hole at planting time. However, don't put fresh manure or chemical fertilizer in the hole with planting. Save that for later to avoid damage to tender root hairs.

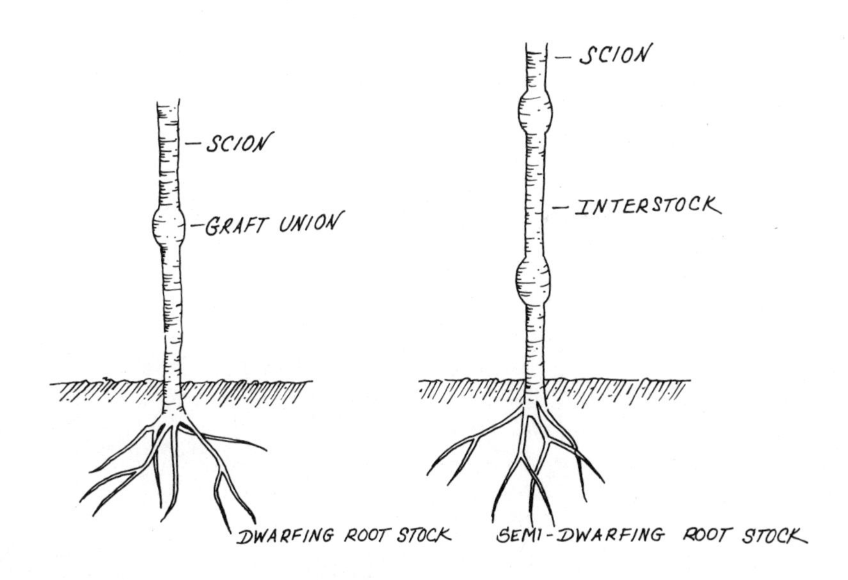

Graft unions tend to produce a slight swelling and change in the bark color or a slight bend in the trunk. Don't cut off or bury graft unions.

Some authorities recommend incorporating manure or fertilizer into the soil. One of the best authorities in America is at Rutgers University in New Jersey. His advice is to add fertilizer later, on top of the ground after planting is completed.

If you have decided to plant dwarf or semidwarf apple or other fruit trees, as most people do today, be certain to keep the graft union *above* the ground. Look at the illustrations in this chapter. They apply to all dwarf or grafted trees and plants, not just apples.

The graft union is indicated by a slight swelling and change in

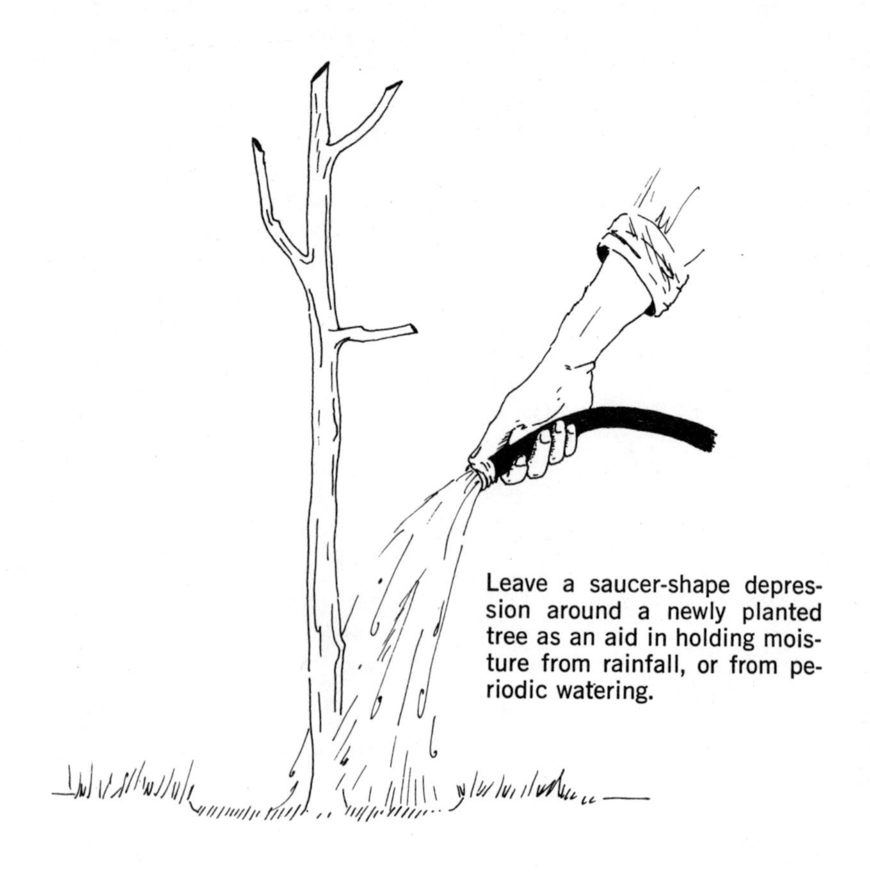

Leave a saucer-shape depression around a newly planted tree as an aid in holding moisture from rainfall, or from periodic watering.

bark color or a slight bend in the trunk near the root. If a middle stem piece has been used for semidwarfing the tree, it will be clearly seen about 12 inches above the root.

Continue to fill the hole with soil, tamping it down with your fist or foot to eliminate air pockets. When it is half full, water well. Then fill to the top and tamp again. Leave a saucer-shaped depression with perhaps a soil dike 15 to 18 inches around the tree. This helps collect rain to encourage new roots to form. Root development occurs when soil temperature is about 45 degrees F.

After the first watering, provide enough water to keep the soil moist at least 6 inches deep, especially during dry periods to get your young tree well started. For winter protection the first year, mound soil slightly around the tree to reduce frost heaving, but keep soil away from the trunk to avoid bark damage. Mulching with straw, compost, peat, and grass clippings also helps protect the tiny, tender new roots that are forming. However, keep mulch away from the trunk to prevent decay from organisms that may be in the mulch.

Mice and rabbits like to nibble on newly planted fruit trees. That famous mouse that girdled an apple tree and produced the first Red Delicious is the only rodent I know whose presence in a garden has been celebrated. All other such creatures are likely to cause trouble.

Use quarter-inch wire hardware cloth wrapped around the trunk 18 inches high. Press it several inches into the ground so burrowing mice can't sneak past. Don't tie it to the trunk. Allow an inch of space, as the illustration indicates, so it doesn't rub against and injure the bark.

FEED APPLE TREES WELL

Many people know they must fertilize their vegetables in order to produce abundant crops. They realize that house plants need nutrients too and often, in their zeal, kill house plants by overfeeding.

For some peculiar reason, lots of folks I know seem to believe that trees just grow all by themselves. After all, they ask, don't trees grow by themselves in forests and woodlands? That's true enough. But when you want to enjoy the bountiful fruits of apple growing, feed you must so your trees will return the favor.

After planting, apply a cup of 10-10-10 fertilizer in a circle about 18 inches out from the trunk. If you scratch it into the soil in that circle, water and rain will then dissolve it properly.

When trees start to grow each spring, scatter 1 to 2 pounds of good balanced garden fertilizer, fairly high in nitrogen, in a circle around each tree. As the trees grow, continue this practice each spring. As tree limbs spread out, they indicate the sideways extent of the root feeding zone of the tree. Root systems grow much more widely underground than you may imagine. On poorer soils a supplement during June may be helpful. Here are some good green-thumb rules for feeding apple trees properly.

Too much fertilizer will cause trees to grow too vigorously. That's bad, strange as it seems. Excess fertilizer will reduce blossoms and produce small quantities of poor-quality fruit. It is better to supply a minimum quantity of fertilizer each spring and supplement it later if necessary.

It is best not to exceed 5 pounds of complete fertilizer for large well-established trees for spring applications. If trees are overvigorous and sprouting sucker shoots and water sprouts, plus extra crisscross branches, reduce your application. A tree in proper vigor will produce about 12 inches of terminal growth by the end of July. If only 2 or 3 inches grow, something is wrong.

As a supplement to the 1 to 2 pounds of balanced 8-8-8 or 10-10-10 each spring, plan to add a quarter-pound of sodium nitrate or the equivalent from some other nitrogenous fertilizer for each year of age of the tree up to 3 pounds maximum per summer.

As you use apple and other fruit trees for part of your total landscape scene, you may prefer some as specimens on lawns. Certain fruit trees—including peach, sour cherry, and plum—grow poorly in sod. Apples can thrive on lawns. However, for all trees in lawns, it does help save time and avoid bruising trunks if you leave a mulch or clean cultivated circle 12 to 24 inches in diameter around each tree. Mulch is preferred, since it also retains soil moisture and provides some minor nutrients as it decays.

PRUNING TO PERFECTION

For some unknown reason, many gardeners—even veterans—overlook the simple practice of pruning. Perhaps they feel guilty about cutting off those happily growing branches. Just as many people are lax in thinning vegetables, so they seem to avoid this other vital need of plants. With fruit trees and berry bushes alike, pruning must be done each year. It not only encourages greater blooming but also produces tastier, better fruit in greater abundance.

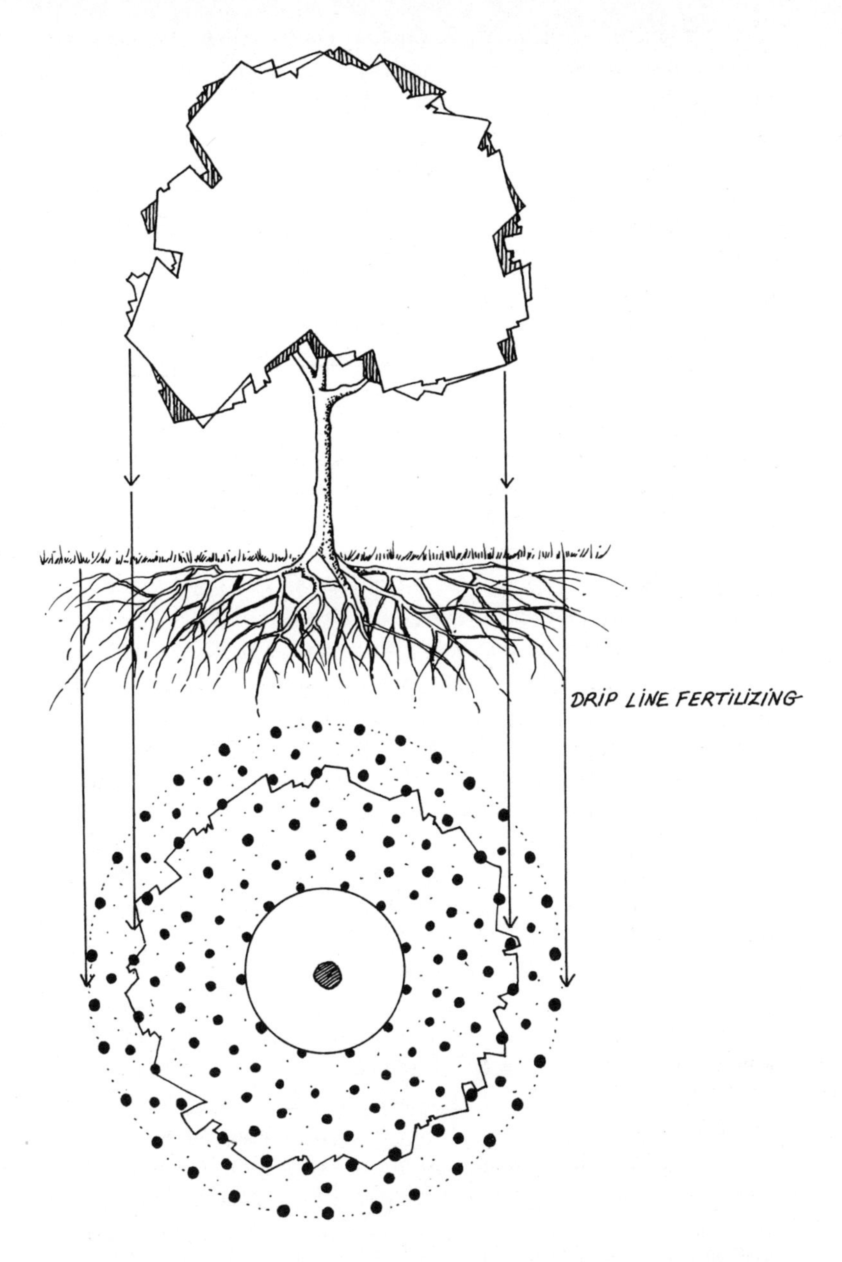

Roots extend much farther than most people realize; so fertilizer ought to be applied in circles out to the "drip line" of the established tree.

In this and other chapters, you'll find easy-to-follow diagrams that show you how to prune your trees from year to year. If you wish to create distinctive designs, follow the tips for espalier training of trees and shrubs. More commonly, you'll want to prune to keep your trees in tidy shapes and thriving within their allotted space of your home grounds.

Since individual trees vary and different varieties naturally assume somewhat different shapes, these pruning tips are basic. You can modify them as trees get older. A branch or two that interfere with a desired view or block out other plants or rub a building or otherwise cause problems can be removed, of course, even if cutting them doesn't fit the exact rules of proper pruning.

The basic framework of your tree is established during pruning in its first and second season's growth. It is best to prune trees in late fall when leaves have fallen, or during mild winter weather. You can see their overall pattern better. Never prune in summer.

One-year or two-year apple and pear trees normally arrive from the nursery as straight, unbranched 4-to-5-foot whips. They hardly look like much. You might think that your hard-earned cash has bought you an ungainly and unsightly tree that can never become what you expected from those lovely color pictures in the catalog. Take heart. It can and will.

At planting time, prune these tall whips to a height of about 3 to 4 feet. (Also prune broken or injured roots to a clean cut before planting.) During their second year at your home, these trees will produce side branches all along and around the main trunk. They may not look capable of doing that, but if you plant, water, and feed them well, they will indeed begin to take a more substantial-looking shape in future years.

At the end of the first season, remove branches to a height of 24 inches. Trees grow from the top and tips of branches. A given point on a trunk, say 24 inches above the ground, remains at that level. Low branches interfere with mowing and other work around the trees. They're not necessary for fruit production.

Remove any branches that form a narrow angle with the trunk. They tend to split when boughs are heavy with fruit and can ruin an otherwise fine tree. Limbs that form about 90-degree angles with the main trunk are strongest. Nature makes them that way. Angles of less than about 45 degrees are weak.

Look at the top of your tree. Remove one of any two branches that tend to divide the tree into a Y shape. Leave only one central leader. If there is damage to the leader, remove it. Trees naturally tend to send up a new leader from one of the topmost branches near the trunk.

If you have bought and planted two-year-old or older trees, prune them according to directions for trees that have completed their first years of growth in your garden. Remove limbs that form weak crotches. Keep only the best branches along the main trunk and a leading branches that is a continuation of the main trunk.

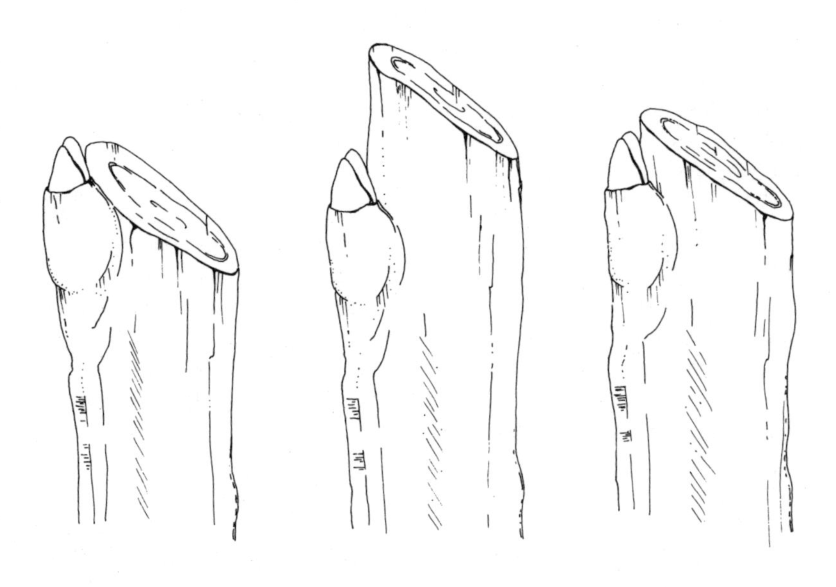

These examples of pruning show, left to right, cuts that are too close to the side bud, too far beyond the side bud, and just the right distance from the bud.

It may seem harsh, but it is really best to prune second-year trees back severely, removing about half their growth. That pruning establishes their proper form and framework for years to come.

After the second season, and until your trees begin to bear, only corrective pruning is required. Don't prune more than necessary during these years. Excess pruning then only encourages long sucker growths in tops of trees. Moderate, careful pruning the third, fourth, and fifth years results in earlier fruiting.

When your tree begins to bear its tasty crop, limbs will bend with the weight. Don't worry. This natural growing process helps open the tree and give it a more spreading form and balance. Annual growth also will be somewhat reduced, so less pruning will be needed.

In future years, thin out branches that rub together. Leave the strongest and best that establish your tree's shape. Remove water shoots and sucker growth. These are the long, spindly shoots that sprout tall and useless from the trunk and along limbs. They serve no useful purpose and bear little if any fruit.

Individual apple-tree varieties differ in growth habits. Adjust your pruning efforts to keep the tree open to air and sun. If storm damage occurs, prune judiciously, removing only the damaged branches so new growth can fill in and rebuild the tree's symmetry.

If you plant hedgerows, remove branches and limbs that grow out in the opposite direction from the row. Keeping apple hedges only 3 feet wide offers a challenge, but they can be kept neatly in line and bearing well for years.

The diagrams in this chapter and in Chapter 19, Pruning Pointers, will help you take those first cuts. After that, prune moderately to maintain the size, shape, and pattern that pleases you and fits your plantscape plan. But do prune. It is a vital aid to nature in stimulating more abundant harvests year after year.

Consider this fact of apple tree life. Fruit wood of a productive tree is continuously weighted down by each year's crops. These limbs never fully regain their height again.

Scaffold branches crowd and depress lower branches. That makes it necessary to continuously remove and thin out some of the other fruit branches. Never fear. New upright growth appears on upper portions of the scaffolds. That's what you want. By cutting out drooping branches and leaving the more upright growing ones you are letting your tree adjust for wood that must be removed. A vigorous tree, well fed and tended, may be twenty to thirty years old. However, the fruiting wood created by your regular pruning efforts is only five to ten years old. In effect, pruning lets you keep your tree young and prolifically productive.

TAKE TIME TO THIN

Each spring you'll see buds awaken and burst forth in bloom. Each bud is capable of being pollinated to produce an apple. Seldom do they all get visited by bees to become pollinated. Some years, how-

A possible fringe benefit of growing apples at home is the ready supply of candidates that could wind up on the teacher's desk.

ever, favorable conditions will produce an overset of fruit. That can weigh down your tree's branches and break them.

When this happens, you might decide to prop up branches. That action is usually foolish. Too abundant a set of fruit most often results in smaller, less desirable apples. It is far better to take time in spring to thin your fruit. This simple practice lets those remaining be nourished fully to produce the plumper, larger, juicier apples on the boughs.

Most fruit trees do have a natural drop of newly set fruits, because of high winds, because the tree "knows" it cannot supply all fruits with sufficient food and water, or for other natural reasons. Nature planned things that way.

If the remaining fruits, after natural drop, are still too plentiful to mature as well as they should, thin them by hand. An excess is anything more than one fruit for each 6 to 8 inches along a branch. Simply pick off tiny fruits within twenty days after trees have bloomed or dropped their petals.

If you don't hand-thin on dwarf trees, especially, they may set far more fruit than they can carry. When they do, you'll find fruits smaller and less tasty at maturity, and your tree may not bloom the next year.

You may find that only scattered limbs on a particular tree need thinning. It's not so big a chore. An hour or less per tree is usually all the time required.

The result of thinning apples on your tree is that it produces (top) fewer but larger fruits than it otherwise would (bottom).

PEST CONTROL POINTERS

Insects enjoy eating apples as much as people do. Codling moths, aphids, and other chewing, sucking, and tunneling pests can plague you if you're not prepared to fight them off. Rather than provide a detailed pest-control program for each particular crop, I have detailed a basic pest-prevention plan in a separate chapter.

New chemicals are being perfected and introduced, and advances are being made in pest control every year. Some pests are more prevalent in certain parts of our country than in others. Control is important, since pests don't damage just a few fruits and leave the remaining ones for you. They can destroy an entire year's crop unless you control them.

Consult Chapter 18 for a basic plan to win your battle with the bugs and blights. Be alert for new products that offer even better pest control as they are introduced through garden centers locally. You can beat the bugs when you know how.

HAPPY HARVESTING

Within a few short years, dwarf trees begin to bear. Standard trees take longer. Some varieties, like *Northern Spy,* may try your patience before they become productive. But when those first fresh, shiny apples ripen, be prepared.

Preharvest dropping is natural. You'll find that some varieties fall more readily than others. As your apples seem plump and ripe, check around the trees. It pays to pick the fruit before it drops and bruises. Apples continue to ripen well off the tree and can be stored in a cool part of your basement or garage. Use those with bruises or blemishes first; don't try to keep them. One bad apple really can spoil a bushel.

Drops can be used for cider. In fact, now that fruit growing has become so popular again, many hardware stores and national department stores with extensive mail-order catalogs offer presses for apples and other fruit. Try making cider if you like. Instructions accompany the presses.

Recipes for apple jelly and for pie, strudel, cake, and other delicious delights abound in magazines and cookbooks. I've included tips for using apples and other fruits in my book *Inflation Fighter's Preserving Guide*.

Once you have enjoyed the first few apples from your trees, you should reread this chapter. Improve the soil around your trees. Feed them well. Be sure they have sufficient water as harvest time approaches. Prune properly every fall and winter. If you follow the simple apple-growing steps, you'll be assured of more and better apples year by year.

5.
Pick Pears for Pleasure

Pears are not as widely grown in home gardens as they deserve to be. Among tree fruits, pears are probably among the best for home planting. They're not troubled seriously by most common diseases or insects that attack other trees and can be grown with very little, if any, spraying. Once established, pears require infrequent pruning to continue producing abundantly.

However, there have been several reasons why pears have not achieved high marks among home fruit growers. Perhaps the biggest reason has been their susceptibility to fire blight. It is a serious affliction that can cause sudden wilting and browning of new growth. Fortunately, improved varieties today are more resistant to this bacterial disease. In addition, new fungicides are available to control it.

Pears also are notoriously poor pollinators, some varieties more so than others. Some varieties require other pear varieties to achieve the necessary pollination to set fruit. For those of us with little area, pears perhaps just don't seem worth the space needed for two or three trees to provide adequate cross-pollination. A third factor may have been the space each standard-size tree required.

Today, none of these factors should bar you from enjoying pears. Fact is, you can grow them quite successfully in many parts of the country. They thrive in various soils and growing conditions. Once established, they require little care.

Interest in pears is reviving, now that fire blight can be controlled both by selecting resistant varieties that are not as susceptible to the disease and by special spray programs that can lick it.

Moreover, pear trees have been successfully reduced in size so they are more suitable for multipurpose landscaping. According to nurseries surveyed recently, demand for dwarf pears is increasing steadily. Pear trees are dwarfed by growing them grafted to a quince root. That's fortunate, since quince is one of the hardiest rootstocks among fruit trees.

In early efforts to dwarf pears, researchers found that certain varieties were not compatible with the quince root. The grafts they attempted didn't grow. Searching for ways to overcome this problem, plant breeders discovered that a compatible interstem of *Old Home* can be used to produce a strong dwarf pear tree. Today, many excellent varieties are being made available. Pears deserve a comeback. They are a truly tasty addition to multipurpose landscapes.

Pears also have another advantage that is gaining them new friends. They respond well to espalier culture. You can train dwarf pears to trellises, create fan-shaped trees, and use them in other distinctive ways.

Earliest records of pears in the United States are traced to Salem, Massachusetts, in the early 1600s. Although fire blight did limit commercial production to far western states where summer rains are fewer, the renewed interest for home growing is encouraging.

Pear tree longevity is amazing. Pears may live and produce abundantly for 100 years or more. *Kieffer* and *Tyson* are two varieties noted for their endurance and continued bearing ability.

Unlike apples, which have variations in flavor and texture but are similar and readily identified as apples even when you're blindfolded, pears are markedly different in flavor. They also have a greater diversity in size, shape, and texture. You can select the small *Seckel* pears or other so-called winter types that keep well and are tasty for preserving or pickling, or the plump *Bartletts* that dribble juice with every bite.

All of the important pear varieties that are available in the United States trace their ancestry to European species. *Kieffer* and *Barger,* however, resulted from crossing European strains with a Japanese pear.

As with apples, there are some 2,000-plus varieties, but only a handful are really suitable for home gardens. No matter. Those that are available are superb indeed and well worth consideration in your fruitful landscape planning.

Pomologists like to say that most pear varieties are self-unfruitful. Translated, that means you need two trees to ensure cross-fertilization so they set and bear fruit. That's fine. Plant two or three or four of the dwarf types and you can enjoy the varied tastes and textures from these different kinds. More people today see this need for more than one pear tree as a bonus rather than a drawback.

Here are some of the better varieties as you go shopping to put pears back in your good gardening life.

The best bet is to stick with those that are known to be resistant to fire blight if you live in the eastern United States, where summer rains and warm weather encourage that particular bacterial disease. *Moonglow, Magness,* and *Seckel* are resistant. They all do well in home gardens.

Moonglow bears early. The trees are vigorous and very upright in growth with large, moderately juicy fruit that is nearly free of grit cells. The fruit is mild-flavored and of good quality for dessert or cooking uses.

Magness bears late on vigorous, spreading trees. Consider whether you prefer upright or spreading growth patterns as you design pears into your plantscale. The fruit is medium-size, oval, and, being tough-skinned, somewhat resistant to insect puncture or decay. The flesh is soft, very juicy, and sweet. This pear is aromatic too. However, the pollen is completely sterile, so you'll need another pear variety to achieve cross-pollination.

Seckel pears mature late in the season. The tree are large, vigorous, and upright with dense tip and some spreading habit. They are amazingly productive, yielding small, symmetrical, smooth fruit. The flesh is white with a slight yellow tinge. It has been described as buttery, very juicy, aromatic, spicy. Seckel pears are known for their excellent quality for dessert or cooking.

Gorham is another good selection. It is a Bartlett-type pear that matures early and keeps in cold storage longer than other varieties. The fruit is bright yellow with a small amount of russet around the stem end. It is hardy and less susceptible to fire blight than Bartlett.

Bosc is a different type of pear. It ripens late, but its quality is deliciously good, especially if it is ripened off the tree. Boscs are large pears with dark yellow undercolor with a veil of fine russet. The flesh is white, tender, and very juicy with tantalizing aroma. The trees develop a somewhat straggly framework of branches and are slow to begin bearing. Once it starts, this variety is very productive. It is a good pollinator.

Bartlett is superior to all varieties both for fresh use and canning. It is large, golden yellow, and highly flavored with lots of juice. The trees grow well and bear early and abundantly. Bartlett is midseason yielding, but moderately susceptible to fire blight.

Clapps Favorite is the best early-ripening pear. It looks similar to Bartlett but with pale lemon-yellow fruit and a bright-pink cheek. Its finely textured flesh is buttery and juicy with a distinctive, delicate flavor unlike Bartlett. The fruit is best when picked "hard ripe" and allowed to fully ripen off the tree. This tasty variety unfortunately is susceptible to fire blight.

Devoe is a late midseason variety that grows on vigorous, productive trees. It has upright growing tendencies with large, long, golden-yellow fruit. The flesh is firm, juicy, and sweet. It is resistant to the blight and needs little maintenance once well rooted.

New York 2480, New York 8760, and *Highland,* also known as *New York 10257,* offer their own delicately flavored pear pleasures. Highland dwarfs well and as a home-garden plant is gaining popularity for ornamental purposes as well as good eating.

Pears enjoy the same growing conditions as apples do. They do prefer some chilling weather to induce their needed dormant period. If they don't get sufficient winter chilling, such as in southern areas, there may be an uneven opening of flowers in spring. That unevenness may create a problem in timing sprays for codling-moth control and interfere with cross-pollination on varieties that require help from nearby other varieties.

On the other hand, most pear varieties will endure winter temperatures well below –20 degrees F without serious injury. *Clapps Favorite, Seckel,* and *Anjou* types are most resistant to cold. *Bartlett* has proved somewhat less so. Pears can withstand higher summer temperatures than apples. In fact, *Bartlett* prefers higher temperatures to reach its peak quality. It fits better in southern areas.

Pears prosper even through droughts and can produce well on a wider variety of soils, including sandier or heavily wet soils, than almost any other tree fruit. For practical purposes and to avoid stressing our pear trees, plant them on soil that is deep, fertile, and well drained. After all, they deserve good growing conditions if you expect them to overcome old-time obstacles and provide you with plenty of pears for your family.

PLANTING POINTERS

Pears and apples not only enjoy the same growing conditions in general, but also should receive the same initial care in planting. Rather than repeat the steps here, I recommend you consult the planting methods outlined for apples. Because pears can live for human generations, it pays to plant them well so their roots reach down and out to get the strongest possible growing start.

Pears do well in western areas of the United States. However, they won't perform satisfactorily on alkali soil or soils that are subject to excess salinity from irrigation water.

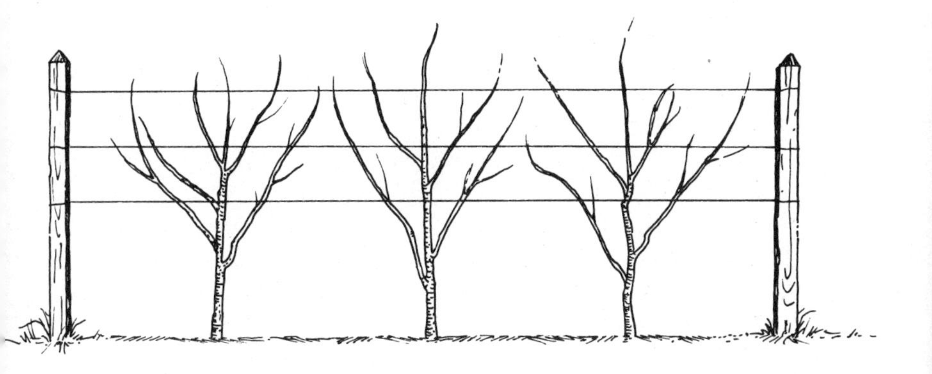

Grouping several pear trees, as beside this wire trellis, produces a balanced effect and provides closeness to insure proper pollination.

Pears can fit nicely in a group at the corner of a yard or in a row along a property line. Grouping pears together produces a balanced planting effect. It also provides the closeness needed to ensure proper pollination when one of the group needs that aid from another.

For hedgerow planting, space standard pear trees 8 feet apart in rows 12 feet apart. Varieties that spread may require up to 14 feet from each other and rows 24 feet apart if you are setting a mini-orchard. With dwarf-rootstock pears, you can space them as close as dwarf apples. For training to espalier distinction, read Chapter 16, Space Sculpturing Is Tasteful Too.

In Europe, a "spindlebush" system combines plantings of pears and apples. That's logical, since they enjoy the same basic growing conditions. We can certainly learn from those accomplished orchardists. They space dwarfed pears on quince rootstock 4 to 8 feet apart and keep trees neatly trimmed. European farmers achieve exceptional yields.

Recently in the United States, *Angers* and *Provence* quince stock has been used to produce dwarf pear trees that are more compact, denser, and more productive. We planted several during the Bicentennial year, among the thirteen new dwarf fruit trees on our property. They have taken root and are thriving better than expected, considering that temperatures Down East in Maine get mighty cold in winter.

You can achieve novel effects by training a pear tree in the espalier method. For more ideas on espalier, see Chapter 16.

PRUNING AND TRAINING TIPS

Young pear trees need some training. They should be pruned to the modified-leader system as you would prune young apple trees. In areas where fire blight is not a serious problem, trim second-year pear trees to three or four main scaffold (side) branches. In the East, where fire blight still exists, and despite the resistance of improved varieties, leave more scaffold branches extending from the trunk, six to eight at first. If you must later prune some infected branches, you still have a reserve that will fill out to form your mature tree.

Most pear varieties grow upright. That is a distinguishing characteristic. They may seem to be sprouting too tall. Resist the urge to top them off, which is called heading back. If top pruning is necessary, do trim off the tallest leaders. But go easy. Too much heading back by cutting the first tall growing tips can encourage over production of more soft terminal shoots. To achieve a more attractively spread appearance, try braces of wood in the crotches. This technique helps limbs spread gracefully. Remove the braces or wedges in the fall to avoid ice breakage during winter.

Once pear trees begin to bear, branches will be naturally weighted into a more spreading appearance. Just as with apples, this normal situation will produce the more balanced shape of pear trees as nature intended.

When trees begin maturing, there is little you need to do except remove damaged branches or prune to keep trees in the desired shape, whether on a specimen plant, in a hedgerow, or on a specially shaped growing pattern.

Fruit buds of pears are similar to those of apples. They'll have five to seven flowers at the terminal of the cluster. Pears are mainly produced from blooms on spurs, rising from the branches. They may continue yielding for six to ten years. On older trees, it is best to thin out new shoots. From time to time you can remove older branches. Pear trees of bearing age should add 15 to 30 inches of growth per year.

As you prune out shoots and older wood occasionally, remaining shoots fill in to become bearing (fruiting) wood in future years. Actually, the simple practice of removing damaged branches and limbs and shoots that cross or tangle and generally opening up the tree every few years is all that pears require. Overpruning bearing pears is unwise. Just a few cuts here and there to keep trees open to air and sun and sprays when pesticides are needed will be sufficient.

Do remove suckers from trunks and water sprouts too, those ungainly tallest shoots on branches. If you discover fire blight, remove the affected area when the tree is dormant. Destroy the affected branches by burning. Fire blight disease prefers to attack young succulent growth, especially water shoots and sucker growth. So remove them when they form.

If fire blight does strike unexpectedly, take steps immediately. The bacteria overwinters in cankers under bark on limbs and large branches. These areas look darker than surrounding healthy bark, They're usually slightly sunken and rough. During warm, moist spring and summer weather, a brownish sticky liquid oozes from them. Wind, rain, and insects carry bacteria to succulent growth on that tree or nearby trees.

The first control step is to locate these cankers and cut them out. Use a sharp knife and remove all the canker to clean wood. Then disinfect the wound with a solution of one part household bleach to nine parts water. Also disinfect your knife and pruning tools with the same solution between cuts. Cover wounds with tree-wound paint, available at garden and farm-supply stores, hardware stores, and even many chain stores these days.

Prune any brown, dead portions of each twig at least 8 inches below the infection. Disinfect the pruning tools after each cut. Remove all pruned material from the area and burn it.

A streptomycin spray helps reduce fire blight spread, according to specialists in the study of this problem. This spray should be used during bloom, at petal fall, and at ten-day intervals if the problem is pressing in your area. However, consult your local pesticide supplier for his formulas and timetable depending on conditions in your locale.

FIRE BLIGHT SPRAY SCHEDULE

For areas that are threatened with fire blight, here's a suggested spray schedule. It was provided by the University of Missouri but applies for other areas. Check the latest recommendations based on recent updated research in your area.

First Spray	When blossom clusters show tinge of pink	100 ppm* streptomycin
Second Spray	7 days after first spray	100 ppm* streptomycin
Third Spray	7 days after last spray	100 ppm* streptomycin
Fourth Spray	7 days after last spray	100 ppm* streptomycin

**Parts per million.* Streptomycin is sold under the various trade names of Agrimycin, Agristrep, and Phytomycin by various manufacturers.

PEAR NUTRITION

Pears respond to fertilizer programs the same as those you would apply to apple trees, adjusted for dwarf to standard-size trees. Because fire blight is a particular problem of pears, take extra care in fertilizing. It is best to go lighter in fertilizing pears than with apples. Instead of 1 to 2 pounds around a newly established tree, reduce that amount almost to half. If suckers or water shoots grow, remove them. There's no sense letting fire blight from other areas find succulent new territory on your home grounds.

Apples, peaches, and plums usually require some hand-thinning to avoid overbearing that results in too many smaller, less flavorful fruits. Pears need relatively little thinning. Bartlett and Bosc varieties may set heavier crops from time to time. If they do have four to five fruits per cluster after petal fall, you can simply reduce that number to two to three. Usually this isn't really necessary. Remember that excessive fruit set can weigh down and snap branches.

Pears, like any other fruit trees, attract pests. Rather than duplicate spray schedules, crop by crop, I have outlined a basic pest-prevention in Chapter 18. Local conditions, whether in fire blight areas or elsewhere, determine the precise materials and timings of pear pesticide-protection applications. County agricultural agents and state extension services publish yearly recommendations based on what problems have been prevalent the year before and new materials that have been introduced. These pest-control plans are free. Just call or write your county agent for details about your specific area.

PEAR-PICKING PLEASURES

Most other deciduous fruits achieve their peaks—their luscious, juiciest, most mouth-watering flavor—when they ripen on the tree or bush. Pears are different. They reach their highest quality when harvested in a slightly underripe stage. If they begin to fall, you may have passed the prime time for picking. Watch your trees carefully as fruits become plump and well colored for their variety.

Then, pick them when they seem full-size for the variety. A better gauge is change of color and firmness. These changes indicate increases in sugar content as pears approach full ripeness.

The ease with which you can separate the stem from the spur is another reliable indication of time to harvest pears. Knowing exactly when to harvest is a trick you'll learn in time. It may not be the same date from year to year, since growing conditions do vary.

In our pear growing over the years, I had this simple rule, called Swenson's pear-picking clue. When no pears have fallen but they look ripe, I pick two. One goes into the refrigerator. The other I taste. After a week, I repeat the test and also sample the first one I put in the refrigerator. Usually I can estimate when pear picking time arrives and harvest the crop for storage and ripening indoors.

Pears do deserve a place in your home plantscape. When you're mowing the grass some fine late-summer or early-fall day, there's nothing quite like reaching up and picking a plump pear.

6. Pretty Is a Peach, an Apricot, or a Nectarine

Peaches may not be native to America, but they certainly have won wide acclaim since they were introduced by early Spanish explorers. They are grown commercially in thirty states and in home gardens in nearly all states. Next to apples, peaches are one of our most popular tree fruits. They, too, have been honored in song and story. "Pretty as a peach," "a complexion like peaches and cream," and "she's a peach" are only a few of the peach tributes that have evolved in our language.

Peaches are perfect for multipurpose landscapes all across America. The trees are naturally small, requiring little space for glorious displays. Today, dwarf varieties let you enjoy the tasty company of peaches even on balconies or in patio planters. Delicious *Bonanza* dwarf peach trees even grow indoors in pots. Peaches are more versatile than you imagine. If you like them, now is the time to consider where you can fit them into your immediate and long-range fruitful landscape. If their fuzz bothers you, consider nectarines. Apricots also fit small spots and plots.

Home-grown peaches offer more delicious eating pleasure than most store-bought ones. The reason is simple. Peaches are especially delicate. They bruise easily, just as ripe tomatoes do. Commercially grown varieties may be fine if allowed to ripen on the tree. However, to withstand the handling, packing, and jarring of shipment over long distances, commercial varieties of peaches have been bred differently. They tend to be tougher-skinned. They also are picked before becoming fully ripe.

You can grow peaches right in your own backyard or on your apartment balcony. Peaches are friendly. They're not difficult to grow either. The varieties you can select offer honest-to-goodness tree-ripe flavor you can't find anywhere else.

The fame of Georgia peaches has spread to all parts of our country.

Dwarf peach trees are small enough to be used in groups, as shown here, or singly and on balconies or in patio planters.

For this reason many people believe that peaches can be grown successfully only in southern areas. That's just not true. Orchardists in Canada grow delicious peaches. Peaches thrive in upper New York State, in Oregon, even in some sheltered valleys along the rocky coast of Maine. To be forthright, I must say it is difficult to grow many varieties in northern areas. However, plant breeders, being the creative talents that they are, have developed excellent varieties that do persist and perform tastefully in northern areas. Fact is, many of our country's best-known peaches were perfected for growing in Michigan. The climate there isn't exactly blessed with southern sunshine.

Although peaches can be grown in many northern areas, these trees and their fruit buds are perhaps the most tender of all fruit trees that can be grown in northern areas. So your site selection is vitally important to give them every chance they can get to please you and satisfy your craving for juicy ripe crops each year.

Peaches love the sun. They can't take late frosts that nip their swelling buds in spring. Severe winter temperatures, below –15 degrees F, will destroy most fruit buds. Sometimes, even lower temperatures can kill buds if they have progressed to certain growth stages when a late cold snap occurs. That can happen after several weeks of mild weather that encourages buds to swell.

When looking for the perfect spot for your peach plantings, consider several vital factors. These points hold true for apricots and nectarines too. Avoid frost pockets, those low-lying spots where cold air accumulates. They are death to peaches and their relatives. Good air and soil drainage will help to maintain the highest temperatures during frosty spring nights and even during wintery cold periods. Usually a peach planting site should be in an area higher than adjacent land. However, if that means exposing peach trees to prevailing winds that dry and freeze the plants, pick another site.

For northern areas you should select cold-hardy varieties. Even in more southerly areas of the mid-Atlantic states, frost-susceptible varieties can be nipped in the bud at times. Odd weather even in Georgia and Florida or California can cancel out large portions of peach crops some years.

The best bet for peaches is a protected, sheltered area of your grounds. It should be warmed by sun by day, shielded from extremes of winds or frosty air settling at night. You can build a wind screen of burlap on posts to guard a young tree against chilling winds. If peaches have a chance to establish strong rootholds on fertile, well-drained soil, they can build up resistance to the cold. A deep sandy loam soil with gravelly clay subsoil is ideal. If your land doesn't have it, you can improve the planting area below and immediately surrounding your intended peach-planting plot. Many homes have been built on poor soils, or the good topsoil has been removed by the developer. Sometimes when a builder backfills around foundations, topsoil is buried by less desirable subsoil. Perhaps only a light covering of loam is spread on the surface to create a quick lawn.

Even if this hasn't occurred on your land, some soils are just nat-

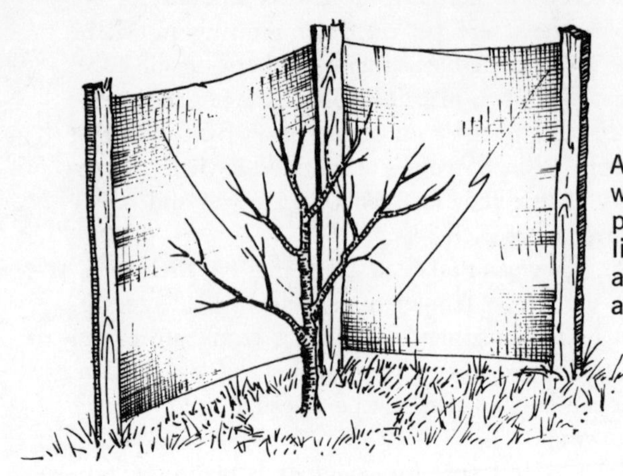

A simple and effective way to protect a young peach tree from chilling blasts is to build a wind screen of posts and burlap.

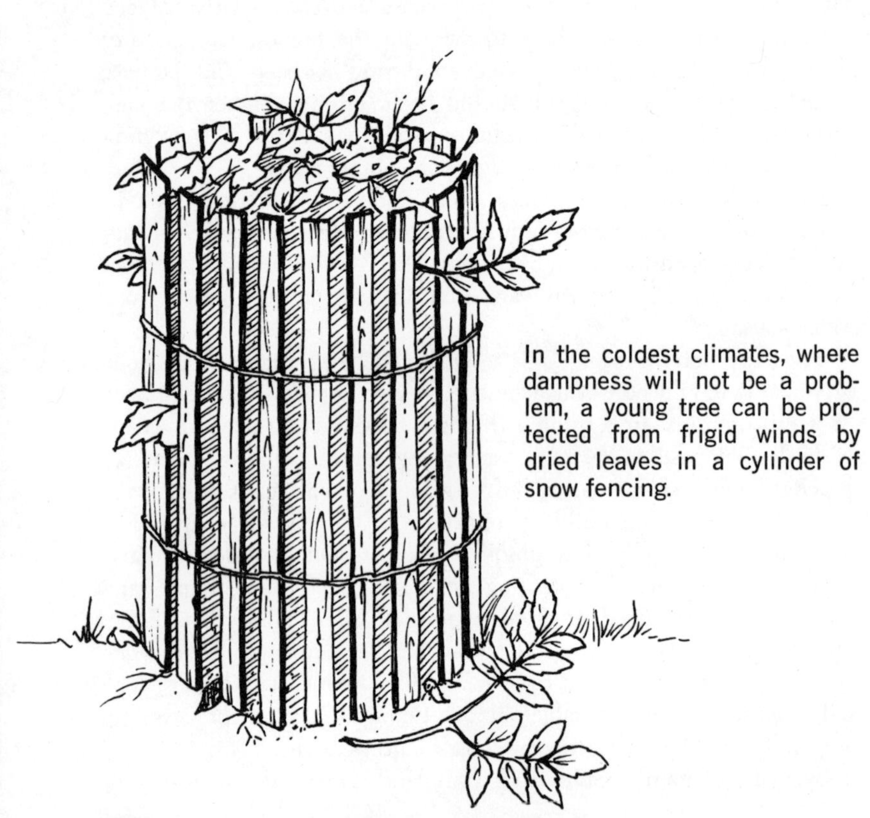

In the coldest climates, where dampness will not be a problem, a young tree can be protected from frigid winds by dried leaves in a cylinder of snow fencing.

urally poorer. No matter what type of soil you have, from heavier clay to sandier type, you can improve it. Details about this are in Chapter 2, Good Earth Basics.

For planting trees, such as peaches, nectarines, or apricots, you can create a much more desirable root-growing area just for the individual trees. Later you can work on improving the surrounding area.

Dig the soil. Discard any that is really poor, rocky, or filled with debris or those wet clay lumps. Then prepare an improved soil mix:

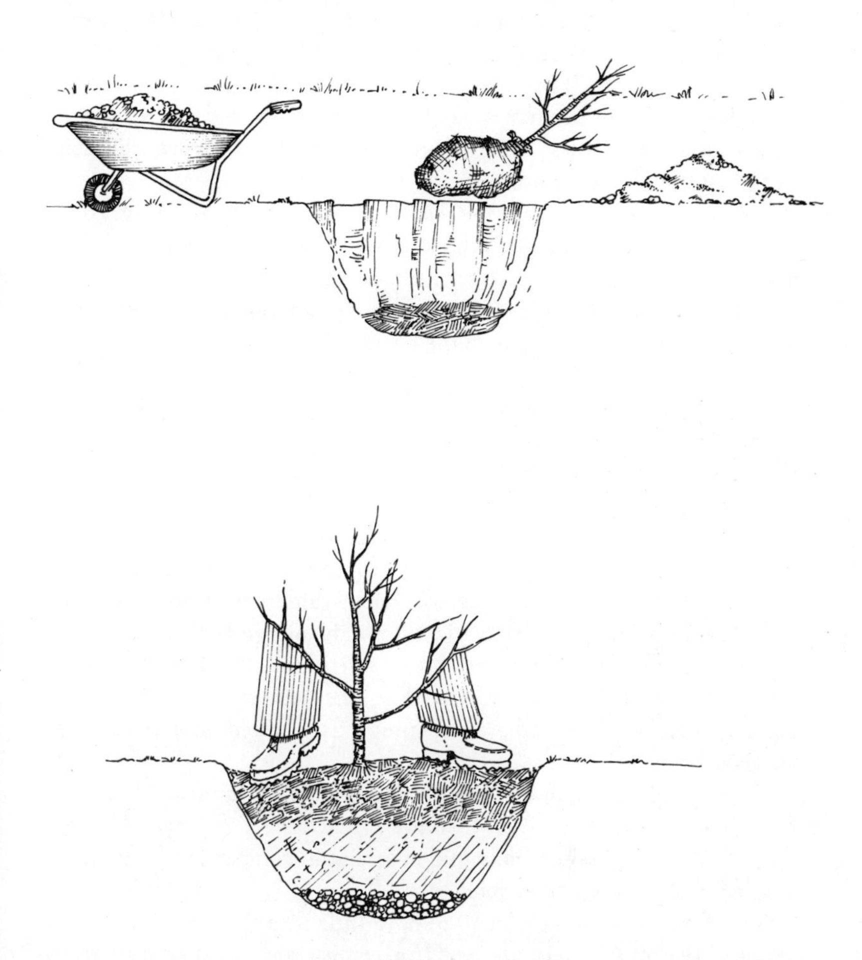

Here's how to produce the desired root-growing conditions for peaches, apricots, nectarines: Gravel is placed in the hole, and the tree is planted in an improved-soil mix.

combine 2 parts good topsoil with 1 part sand and 1 part leafmold, composted humus, or peat moss. Composted humus is best. Mix it thoroughly.

Next place a layer of gravel, about 2 inches deep, in the bottom of the hole, below the depth the tree roots will reach initially. Put well-rotted manure and topsoil, mixed together, on the gravel bed, about 3 or 4 inches deep. Next add a layer of the improved soil mix you've prepared. Then begin the planting process. Spread the roots naturally without bending or crowding. Add the improved soil mixture. Tamp it down. Water well to settle the soil. Add more of the improved topsoil. Tamp that down too. Leave a saucer-shaped depression around the tree or shrub. That will collect rain and direct water from sprinkler or hose to the root area of your newly planted tree or shrub. The illustration shows how to prepare the site, improve the soil, and plant any tree or shrub well.

It may not seem necessary to take these extra steps in planting. After all, you may reason, you and others have planted trees before and they have grown fairly well. That's true. But this extra care for valuable fruit trees and bushes gives them an underground environment that they much prefer, and you reap the results.

PEACHY VARIETIES

New and better varieties of peaches are being introduced very rapidly. Plant breeders are responding to the increased demand for peaches by commercial growers and home gardeners as well. To provide hardier peaches, new varieties are being grafted on *Siberian* seedling rootstocks. These consistently harden off, that is prepare themselves for cold weather, earlier in the fall than trees on more common stocks. This earlier hardening has resulted in much better survival.

For example, work at the Geneva Experiment Station in New York State revealed that trees of *Redhaven* peaches on *Siberian C* stock carried full crops and had low bud mortality in comparison to the same Redhaven variety on other rootstocks.

This Siberian rootstock seems to have little dwarfing effect, but even standard-size peach trees are small in comparison to other fruit trees. They'll fit nicely into most home landscapes. Because of its superior climatic adaptation, fruit specialists recommend use of Siberian C rootstock or similar hardiness-transmitting types for peaches grown in northern areas.

Dwarf peach trees are listed in most nursery catalogs. They are usually budded or grafted on *Sand Cherry* roots. This means they tend to be short-lived. Some nurseries produce dwarf peaches grafted on *Nanking Cherry* rootstock. These are quite dwarfed and highly productive. They seem, judging from trials in various areas of the country, to be longer-lived than dwarfs created on Sand Cherry roots.

Whether you prefer the tiny dwarf peaches that may mature only 3 to 5 feet tall or the standard size, which may be 8 to 15 feet high, the variety characteristics of the bearing wood are what really count. Here are some recommended ones. After peaches, you'll find recommended varieties for nectarines and apricots.

By selecting several different varieties of each fruit you can really stretch that tasty harvest season. Some trees bear very early, others midseason to late in the season. Very early types include Candor, Collins, Brighton, Harbinger, and Garnet Beauty.

Early varieties include Prairie Dawn, Reliance, Redhaven, and Raritan Rose.

For midseason you have a choice of Triogem, Eden, Canadian Harmony, Glohaven, Vanity, and many others.

You have the option, too, of planting some late-ripening varieties. Jefferson and Tyler are two.

Candor ripens early with attractive oval red peaches. The flesh is medium-firm, semicling, juicy, and sweet. These peaches are nonbrowning and well liked for this fact. Trees are vigorous and productive, and seem medium-hardy.

Collins ripens about the same time, quite early. The medium-size peaches mature six or more weeks ahead of *Elberta,* one of the best-known commercial varieties. Collins is early, bright red with yellow flesh. It is semiclingstone. Trees are hardy, vigorous, and productive.

Clingstone, by the way, means that the pits cling to the flesh. *Freestone* means that the pits readily separate from the flesh. For home gardens, either kind is useful, since these processing definitions really have no special meaning unless you run a peach-canning factory or wish to can most of your crop.

Brighton is a high-quality, yellow-fleshed peach with roundish, uniformly medium-size fruit. It also is bright red on a yellow background. The flesh is medium-firm and semicling, juicy and sweet. Trees are vigorous and productive.

Reliance is one of the most winter-hardy varieties. It takes severe cold. The fruit is medium-size, yellow-fleshed, and of good quality. It keeps well and is dandy for eating or preserving. Trees are highly productive and hardy in colder areas.

Red Haven is a superior variety with good winter hardiness too. The flesh is firm, finely textured, and of very good quality. It is versatile for eating, freezing, and canning and is prolific. You should thin Red Haven or it overbears.

Canadian Harmony is large, highly colored, yellow-fleshed, firm, juicy, and nondarkening. Trees are vigorous, productive, and quite hardy.

Among late varieties preferred by many home gardeners, *Jefferson* is a large, yellow-fleshed freestone peach. It has red and orange-yellow skin with a firm, flavorful texture. Trees are vigorous and produce heavy fruit set, which requires thinning.

Tyler ripens late on vigorous trees. The fruits are firm, flavorful, and freestone. This, too, may require hand-thinning since it tends to bear heavily.

NECTARINES

If you don't like fuzz on your peaches, try nectarines. Their smooth skin has made them more appealing to anyone who dislikes the fuzzy-wuzzy feel of peaches. Don't laugh. Many people still peel peaches rather than eat them skin and all.

Nectarines are somewhat more susceptible to brown rot than peaches are. However, this problem is quite easily controlled with an effective combination pesticide and fungicide application program.

Smooth skin and delicious flavor make nectarines an increasingly popular fresh fruit for home landscapes. Their culture is identical to that of peaches. Nectarines have an advantage: varieties are usually self-fruitful. In other words, they pollinate themselves readily. Many peaches don't, so you may need several trees. There are many more nectarines than space permits me to list here, but the following ones are good.

Lexington is a productive, medium-size, yellow-fleshed variety. The flavor is sweet. Trees are hardy and resistant to spring frosts.

Cherokee is a fine yellow-fleshed nectarine. The fruit is large and bright-colored with firm, juicy flesh and semicling habit. Trees are productive and medium-hardy, making this variety better for southern areas.

Tiger is a Stark exclusive. It is exceptionally winter-hardy, resistant to brown rot, and a fine freestone variety of good quality. Trees are productive, but may need some hand-thinning.

Redgold is a very high quality, beautifully colored, and firm-fleshed variety. It is winter and spring bud-hardy too.

Pocahontas is a very early, yellow-fleshed semicling nectarine. The flesh is juicy and sweet. Trees are productive but medium-hardy in northern areas.

APRICOTS ARE TASTY TOO

Apricots prefer well-drained, light- to medium-textured soils of reasonable fertility. They bloom early, so a site with good air drainage is important to avoid spring frost damage. Most apricots are self-fruitful, but it is wise to plant two varieties to ensure maximum fruit set. Since these fruits do tend to suffer in harshly cold areas, they are best grown in parts of the country that have moderate temperatures. They can be grown in northerly climates, but extra efforts should be taken to protect apricots from winter winds.

As with peaches and nectarines, many varieties are available. Catalogs will advise you which are best for what areas. Some are more hardy than others.

Alfred is a productive, good-quality apricot that thrives in New York–type climates. The fruit ripens late in July and is medium-size and bright orange with a sweet, rich flavor to its firm, juicy flesh.

Goldcot is a winner according to Michigan Experiment Station researchers after thousands of field tests. The fruit is nearly round, the flesh medium orange with a fine texture. Trees are exceptionally strong, above average in winter hardiness, and self-fruitful. Heavy fruit set may require hand-thinning.

Sungold is extra hardy for trying climates too. It produces heavily with freestone fruit that ripens in mid- to late season.

Moonhold bears earlier but is not as tasty as Sungold. Moonhold produces heavy crops with a sweet taste and is good for fresh use and jam or all purpose.

Stella is a hardy Russian-type apricot that is very cold-resistant. The fruit is medium-size, golden in color, freestone, and delicious. It thrives where peaches can be grown.

As you shop for peach, nectarine, and apricot trees, keep in mind that they are less tolerant of severe conditions than apples or pears, as well as susceptible to sharp variations in weather. Apricots especially may be a gamble even in sites just right for peaches. The apricot blooms so early it can be nipped in the bud by late spring frosts.

PEACH PLANTING TIPS

Peaches prefer spring planting, as soon as soil can be prepared. It pays, of course, to get the site you have selected ready the previous fall. However, young peach trees are more tender than other fruit trees. Giving them the advantage of a spring start and a full season's growth to set their roots well and acclimate themselves to their new home is desirable.

Peach trees enjoy the same tender loving care you would give to other fruit trees at planting time. Perhaps some extra care is warranted. If the trees you buy are bare root, be sure to keep them moist from the moment of their arrival to the time you set them in the soil by heeling in or planting them.

Soaking peach, nectarine, and apricot trees in water for several hours before planting is a good practice. You should also keep roots moist with wet burlap if you have removed them from a pail of water and delayed the actual planting for any reason. With peaches, as with their close relatives the nectarines and apricots, soaking roots before planting is beneficial.

Plant them as you would other valuable fruit trees. Detailed step-by-step directions are outlined in Chapter 4, Apples Are Appealing, under planting tips. Since peach trees and their cousins are tender, never use manure or fertilizer in the soil mix or added into the hole as you plant the trees. It can be lethal to them. Hold off on fertilizing these trees until several weeks after they have been planted. If the soil is reasonably fertile and you provided an improved mixture during planting, there is no need to apply fertilizer the first year. However, if you did not improve the soil, a cupful or two of 10-6-4 spread in a ring 18 inches around the trunk of newly planted trees can be scratched into the soil two to three weeks after planting.

When you plant, be certain that the bud union is 1 to 3 inches *below* the soil level. That's right. This is different from the usual procedure for other fruit trees. Peaches are more tender. If suckers arise from the ground, just prune them away. Water well twice each week if no rain falls for several weeks. Peaches need and deserve a good start if they are to serve you well in the years ahead.

Once your peaches, nectarines, or apricots are well rooted, it's time to satisfy their growing needs. Nitrogen is the most important element for them and very often the only nutrient they need. Too little nitrogen can result in low yields, poor fruit size, and excessive cold injury. Too much plant food may cause overly rapid growth, poor fruit color, and excessive cold injury.

The amount of nitrogen that peach trees need is based on growth and performance of the individual tree. You should expect them to make 18 to 24 inches of new terminal growth annually. In mature, bearing trees, at least 12 inches is optimum to maintain good vigor. Unlike other fruit trees, peaches and their relatives must be catered to individually. Get out a ruler. Check their growth. Watch bloom, fruit set, and harvest yields. Usually a bearing, mature tree needs about one pound of nitrogen per year for best results. You can apply that in a circle scratched into the soil around the tree 24 to 36 inches from the trunk. Peaches and their friends don't like to compete with sod or grass of any kind. It is best to mulch beneath peaches or clean-cultivate to keep weeds and grasses from interfering with their performance.

Once trees have begun bearing, a 10-6-4 or similar garden fertilizer, high in nitrogen, will usually keep them well satisfied.

In experiments by peach-growing experts, various fertilizing programs were tested over a period of years. Fertilizer rates per tree of bearing size to six years were 1 pound of 16 percent nitrogen or ½ pound of 33 percent nitrogen, 1 pound of 20 percent superphosphate, and ¼ pound of muriate of potash.

Nitrogen fertilizer resulted in an increased tree size. This increase was achieved whether nitrogen was used alone or in combination with phosphorus or potash or both.

Tree growth was not significantly improved by phosphorus or potash. Trees receiving high nitrogen and smaller amounts of the other ingredients did produce about three times as much fruit as trees that received no plant food.

Peach trees are naturally smaller than other types of fruit trees. Dwarfed varieties may be grown successfully in large tubs.

If you try this unusual idea of a potted patio peach, give special attention to planting. A gravel layer in whatever container you use is necessary. Then add the improved soil mixture mentioned earlier. Dwarf peach trees in restricted growing situations need extra attention to their habitat. Be careful, too, with the amount of fertilizer you apply to dwarf peach trees, especially in containers. Too often we all tend to think that if a little plant food is good, more will be that much better. That misguided thinking leads to problems. In containers and on dwarf trees, go light with the fertilizer. A little goes much further than you realize. Read and follow directions on the fertilizer container.

PEACHY PRUNING POINTERS

Dwarf peaches need little care except occasional clipping here and there to maintain their lovely diminutive shape and form.

Training standard peach, nectarine, and apricot trees, however, does require special care when they are young. A tree that is correctly trained will usually live years longer than one left to choke itself by overbranching. The trained tree will resist winter injury to the trunk and main scaffold branches and be able to carry heavier fruit loads with less limb breakage.

Peach, nectarine, and apricot trees are naturally designed for open-center growing. They welcome sun and air into their interiors.

Here's a step-by-step guide from the ground up after planting.

One-year-old peach trees will arrive from mail-order nurseries usually as a branched whip. Often side branches are weak and too small for framework branches. Cut these back to spurs leaving three or four buds on each. The tree itself should be cut back to about 36 inches above the ground.

After the first season, remove all side branches that form a narrow angle of less than 45 degrees with the trunk. Remove any branches that are only a few inches above the ground. They will be too low to produce any fruit and will just hinder regular care.

Also cut off one of any two limbs of approximately equal size that tend to divide the tree into a Y shape. Remove suckers or even strong branches that fill in and shade the center. Peaches, nectarines, and apricots prefer an open tree growth to achieve the fullest, ripest perfection from proper sun and air circulation. To keep your trees well balanced, cut back stronger framework branches slightly. You may allow a central leader similar to pruning and shaping apple trees, but open-center peach and related trees are much better.

After the second season, continue pruning to develop an open-center, spreading, actually bowl-shaped tree. Remove any limbs that tend to grow up through or across the center. Continue to cut away suckers growing straight up. However, do retain most of the other one-year growth throughout the tree.

After two or three years, a well-grown tree will have a trunk 4 to 6 inches in circumference and a good supply of fruit buds. Pruning now takes extra care. Moderate pruning will result in production of as much as a bushel of peaches the third summer. Severe pruning will reduce or may eliminate the third summer's crop.

After the third year, peaches should produce substantial crops annually. They may live twenty to thirty-five years, depending on the care they receive. Continue to prune to achieve that open center. Peaches are produced on wood that grew the previous season. That's why continued pruning is always necessary to stimulate that new wood which will bear well for you the next year. Depending on your landscape use, you can train peaches moderately. However, if they are primarily as specimens or part of a home mini-orchard, try to keep them about 12 feet high. That size is hardier and easier to maintain, prune, and harvest.

After several years' growth, peach trees can develop two or three scaffold branches of great strength. Two are better for overall shape. Keep pruning to a minimum after the first few years, just enough to maintain shape, remove dead or damaged branches, and stimulate next year's fruiting wood.

PEST CONTROL

Brown rot and powdery mildew may trouble your peach trees and their relatives. Fortunately, modern fungicides can avoid these problems. Insects, too, can attack at times. Details on pest control, based on a sound pest-prevention plan, are included in Chapter 18, which covers spray schedules for all fruits.

Thinning nearly every year may be necessary to ensure a crop of plump, full-sized peaches. Otherwise trees may overbear, break branches, and set too many fruits that won't be as nice. Handthinning is easy. Details about this simple practice are included in Chapter 4 under thinning.

Consider that it takes 380 peaches 2 inches in diameter to fill a bushel, but only 190 of 2½-inch peaches to fill the same bushel. Wouldn't you rather have those bigger, tastier ones? You'll end up with the same amount of bushels, actually.

Thin peaches so they are spaced 2 to 8 inches apart on the tree. Proper thinning has also increased winter hardiness, according to experiment-station reports at agricultural colleges.

WHEN IS RIPE?

Mature and ripe are often used interchangeably by farmers and orchardists. Actually, these words describe distinct and separate processes. Maturity involves fullness of growth and indicates completeness of development of the peaches, nectarines, and apricots. Ripeness takes place after fruit is mature. It involves the softening of the flesh and the development of juiciness and flavor. This stage may occur before or after fruit is taken from the tree, since peaches also ripen after picking. The best way to tell when your peaches are near their peak maturity is to watch the ground color—that is, the undercolor of the skin. It may be greenish or greenish yellow at first. As ripeness approaches, it will become more yellowish or orange overall. The red color may brighten but is never as good an indicator of ripeness as the ground color. Feel, of course, is helpful, but pinching peaches does bruise them. Remember that not all peaches mature at the same time. Selective picking is in order, perhaps three or more times. Take only the largest, most mature, and fully colored ones each time.

If you plant several peach trees and have done your homework on early- to late-season maturing types, you can enjoy the best flavor that peaches have to offer from early summer right through almost until the fall.

The planting, thinning, pruning, and harvesting methods for nectarines and apricots are similar to those for peaches. However, apricots do bloom earlier. If you wish to try them, be certain to insist on the hardiest types. These delicious fruits are especially tender. Winters in northern areas can thwart even your best efforts to grow them.

Look around your land. Pick that fertile, well-drained area. Improve the soil, and get to planting. Peach picking time in the home landscape is just only a few years away on those quick-maturing fruit trees.

7. Perfect Plum Pleasure

Little Jack Horner wasn't the only one who loved plums. Pioneers and the millions who came after them to settle our vast country brought plums to plant from their homelands. They found the climate and soils ideal here. Fact is, plums can be grown more easily in more parts of our country than practically any other tree fruit.

Plums offer a greater selection of species and varieties adapted to the different climate and soil conditions that are found across America. Some can tolerate extreme cold areas; others prefer and thrive in more temperate climates; still others grow well in warmer regions. Being relatively small in size, plum trees can fit into a variety of garden situations. They bloom beautifully in spring and provide foliage displays in summer and fruit to save and savor all year round. As you scan your grounds this year, pick a spot or two for plum trees. They, too, can be trained in distinctive patterns, as indicated in Chapter 16, Space Sculpturing. Even without much extra attention, they'll provide cascades of bloom and bountiful harvests of succulent fruit from relatively small space.

More than 2,000 varieties of plums have been grown in the United States, according to best estimates by plant scientists. Today, your choice of types is wide and flavorful. The most important type is descended from European plums. It is moderately vigorous stock with thick leaves that have glossy green sheen above and paler green color below. Fruit grows on spurs and varies in size, color, and shape, with distinctive differences in flesh color and flavor too.

The plum family from European ancestors has five basic groups. One is the *prune* group, which can be picked and dried with its pit intact. The *Green Gage* group is more easily identified by its round fruit with green, yellow, or reddish flesh, sweet, tender, and juicy. The *Yellow Egg* group is primarily a commercial canning type. The *Im-*

peratrice group includes most of the blue-colored plums. These trees typically bear heavily, producing medium-size, oval-shaped fruit with firm flesh. They are not as flavorful as others. The *Lombard* group is similar to the Imperatrice type but is red in color, usually smaller, and somewhat lower in quality.

American plums are another family, native to America. They include species suitable for fresh use or cooking. Plant breeders have found them useful as rootstock on which more desired varieties can be grafted to achieve a hardiness for more severe areas.

Japanese plums are typically early bloomers. For this reason they may be susceptible to frost. Many are as winter hardy as peaches and can be grown under the same range of conditions. Others won't survive far-northern climates. This type of plum is distinguished by its larger size and tendency to be heart-shaped. It may be red, yellow, or a blend between. Japanese plums also tend to be more colorful at blooming time because they have a more abundant flowering habit.

Plums may be self-pollinating or require cross-pollination from a nearby similar plum variety. Nurseries can advise on this situation, so do inquire. There's not much sense in planting just one that you think will bear, waiting several years, and then realizing another tree is needed for cross-pollination.

As you look around your land, estimate spacing for plums as you would for peaches. At maturity they'll take the same space as standard peach trees. You can, with care, keep them in bounds with proper pruning. However, since they are reasonably small, it's best to give them the room they need for their most abundant performance.

Here's a brief list of some of the better plums you can select. Mail-order nurseries offer others that may be better suited to specific growing conditions in certain areas. If you have doubts, it pays to ask. After all, it's your cash that pays for the trees. And when you put the work into planting, tending, and cultivating them, you have every right to know they'll prosper in your area.

EUROPEAN PLUMS

The European plums comprise the most important and widest selection of types and some plums of the best quality available. Many do require cross-pollination. Even varieties classed as self-fruitful may produce much better and more abundant crops when cross-pollination from other trees is provided.

DeMontfort is an old French blue plum. It bears in August, is medium-size and roundish oval, and has dark purple fruit that is juicy, sweet, and rich. It is highly regarded as a freestone type.

French Damson is one of the largest Damson plums. It is vigorous and productive. The fruit measures 1¼ to 1½ inches in diameter and is delightful for making plum preserves.

Green Gage has been cultivated in Europe for hundreds of years. You can grow it here quite easily. The fruit is medium-size, yellowish-green, mottled with red. In Europe, it is considered the ideal dessert plum, tender, juicy, rich in melting fresh flavor. If space is limited, this perhaps should be first choice.

For early plums, try *Leaton Gage*. It has high quality and is fairly productive. Trees are vigorous growers.

Oullins is an early-ripening Gage type, one of the most attractive and largest available. *Reine Red* is a Red Gage that ripens late.

Oneide is a late, reddish-black prune-type plum. It has proved consistently productive and is quite self-fruitful.

Stanley is a prune-type plum variety. Trees are hardy and vigorous and produce large crops annually. The flesh is greenish yellow, juicy, sweet and firm, good for eating or cooking.

Yakima is a vigorous, upright-growing variety. The fruit is very large, oblong, and a bright mahogany-red. The flesh is tender, firm, sweet, and yellow-colored; it's fine for dessert use.

Vision is a highly vigorous and productive plum. Trees are spreading. The fruit is large and blue-skinned with yellow flesh. It is high-quality and freestone.

If you live in colder climates, here are some plums that have that necessary cold-hardiness.

Mount Royal is a prune type, blue in color. It makes a great dessert, jam, or preserve. It has survived extreme cold in test growing conditions of Canada.

Purple Heart is a plum with superior winter-hardiness developed in New Hampshire. It has deep red fruits of medium size with sweet red flesh. The tree is low and spreading and fits beautifully into landscape scenes.

JAPANESE PLUMS

Japanese plums have been favored by many plum fanciers for their taste and large size. If you prefer this type, remember that nearly all

Japanese varieties require cross-pollination. At least two different ones should be planted to ensure that they bear a crop.

Formosa is recommended for its large, attractive oval fruits. Trees bear biennially, yielding greenish-yellow fruit overlaid with red. The flesh is firm, juicy, sweet.

Santa Rosa is one of Luther Burbank's more noteworthy developments. It is a prolific bearer with large, attractive fruits.

Burbank is a round plum that ripens early. Trees are hardy, both themselves and their buds. It produces heavily, yielding medium to large fruits that are juicily sweet. These trees are low-growing and somewhat drooping, with a tendency to be distinctly flat-topped. They require cross-pollination.

Mariposa plum is large and purplish red with blood-red flesh that is sweet and nicely juicy. It is very winter-hardy but also needs cross-pollination.

Shiro bears early, rewarding you with bright yellow plums with bluish bloom. The flesh is light yellow, juicy, sweet, and favored for dessert or cooking. Trees are upright and spreading, large and vigorous.

Burmosa is another early bearer, moderately productive, and vigorous with large, round to oval fruit, greenish to amber-yellow skin. The flesh is sweet, melting-mild in flavor, and of good quality for eating fresh.

PLANTING POINTERS

Plums don't grow true from their pits. Instead, nurseries must propagate desired varieties by budding seedlings onto hardier rootstock in the nursery, as they do for peaches. Plums are grown mainly on Myrobalan stocks, but some plant breeders do produce them on seedlings of peach or Japanese apricots.

The *Myrobalan* stock has advantages. It is tolerant of poor soil conditions and adapts to a wide range of soils, from sandy to clay loams. It also has proved hardy and deep-rooted and has excellent longevity.

Plums can be dwarfed. Some nurseries do offer this choice, although the natural small size of plums may make dwarfing unnecessary. Breeders have used the *Western Sand Cherry* as a dwarfing stock for varieties of European plums. When it's used for Japanese varieties, the resulting trees are very small. That quality fits them well if space is severely limited for the two or three needed to ensure proper pollination with Japanese types.

In some areas, an underground microscopic eelworm has caused problems in plum orchards and home gardens too. Today, plum and peach stocks are available that are resistant to root-knot nematode. Soil fumigation before planting, in areas that have this soil nematode problem, is advisable. Your county agricultural agent most likely knows whether that would be necessary in your locale. It's tricky to do, so if there is a problem where you live, particularly in southern areas, perhaps a switch to other fruit trees is best.

Space plums the same as peach trees. Many nurseries offer started plums, since grafting or budding time in the nursery lets several laterals form on the trunk. However, you also may get plums as one-year-old seedlings that have the same unbranched whiplike appearance as young apple and pear trees.

Here's a general training rule. Plums that have an upright-growing habit such as Stanley, Santa Rosa, and similar varieties should be trained to a modified leader system, much as you train young apple trees. The same holds true for many of the European types, which tend to be larger and more upright.

The Japanese-type plums—with their more spreading, lower growth—are best trained to the open system, as you would peach trees. Rather than repeat all the details for apple-tree and peach-tree pruning, I'll refer you to the pruning pointers in Chapters 4 and 6.

Japanese varieties of plums seem to do best when pruned to four to five main scaffold branches. They also tend to produce more unwanted lateral shoots and water sprouts. That just means you'll need to pay more attention to cutting back excess lateral growth and removing sucker shoots or water sprouts.

Perhaps a few additional pointers are in order here. Mature plum trees bear fruit laterally on one-year-old wood as well as on vigorous spurs of older wood. You should expect shoot growth of about 12 to 24 inches on young trees. Mature trees will produce about 10 inches of new growth each year.

Japanese plums tend to overbear. You can control this tendency with somewhat heavier pruning as well as thinning of extra young fruits as they form.

Plums have another unusual habit. They tend to drop young fruit a short time after setting it. That's normal, so don't become alarmed. Do, however, check for insect damage that may cause excessive fall of young fruit. A little fruit drop is helpful. It is the tree's natural way of thinning itself. After drop occurs, check the fruit set. If there still are too many tiny plums too close together, hand-thin them as you would for peaches.

Fertilizing is necessary in order to maintain the vigor and productivity of your plum trees. Since plums prosper with the same care on the same type of soil and cultivation as peaches, you should reread that section of Chapter 6.

Plum trees, especially the Japanese types, respond cheerfully to espalier training. You can tie them into distinctive shapes and adjust your pruning efforts to create unusual forms and fancies with them. Chapter 16 provides those definitive details.

Pest control for such uglies as the plum curculio and other insects that may attack your plums is well treated in its proper place. For both insect and disease-control recommendations, consult Chapter 18.

HARVESTING HINTS

Plum-picking time varies with varieties, which is only natural. Some trees bear early, others mid to late in the season. When plums seem to be well filled out, plump, and mature, begin to check them every day or so. When to pick depends on the use you plan to make of your plums.

For canning, preserving, and other cooking purposes, it is best to pick plums when they are firm-ripe. They will retain their shape and perform better in cooking and canning at that stage of slight underripeness.

However, if you plan on jams, preserves, and jellies, wait until plums are fully ripe. They will then give you higher natural sugar content for a sweeter end product. For fresh use, the riper the better. Just bite into one or two a day as they approach the ripest appearance on the tree. Your taste will tell you when they're truly ready.

Color is another help in judging ripeness. Most varieties will change markedly within a week before they're ready. With green-colored plums, you'll notice the green gives way to yellowish green before adding that pinkish blush some varieties attain.

Blue and purple plums change gradually from greenish blue to shades increasingly darker until fully ripe. Red plums also change within ten days from their light tints to the darker color of fully mature and ripe fruit for the individual varieties.

Plums promise lots of pleasure. They're hardy and vigorous. And despite the need of some varieties to have neighboring trees for cross-pollination, they are abundantly fruitful. Plums do taste different from other tree fruits. Give yourself a variety of taste treats from your garden areas. Plant plums for that added measure of plum tasty pleasure.

8. The Charm of Cherries

Cherries are a favorite all around the world. Some are sweet, others sour. Both have their place in your home landscape plans.

Perhaps George Washington didn't like cherries. Nobody really asked him *why* he cut down the tree, just whether he did or not, right? Nevertheless, cherries fortunately survived that early bit of axmanship. Today, sweet and tart cherries remain popular. Fact is, commercial acreage of them in the United States has climbed dramatically in the past twenty years.

Most of the cultivated cherries today are derived from two species. One is the sour cherry, from which both the light- and dark-colored varieties have developed. The sweet cherry is another type and holds its own in popularity even though sour types make better pies. Most of the cherry varieties suitable for growing in the United States are descendants of stock from France, England, Holland, and Germany. Since cherries do produce more or less true to type from seed, it is difficult to trace exact lineage. However, fruit breeders are providing even better varieties worth trying in home landscape today.

Cherry trees grow to medium height, between the size of standard apples and standard peaches. For that reason, they fit neatly into many home plantings. They also bloom earlier than most other fruits. Their more graceful shape and abundance at harvest time have made them favorites through the years.

A complete list of these delicious tiny fruits would take pages. More than a thousand varieties have been grown at one time or another in this country.

Some types are well known, most likely for their popularity fresh or for pie making. Bing and Montmorency are two. Others are less well known, yet even better for home growing than commercially suited varieties.

Sweet-cherry trees tend to grow larger and with upright growth pattern, much like apples. Sour-cherry trees mature more in a spreading shape and closer to the size of peach trees. In fact, pruning techniques for each type parallel those for apples and peaches. More on that, of course, later.

Sweet cherries have one disadvantage. They are, as scientists phrase it, self-unfruitful. For simplicity, that means you should plant two compatible varieties near each other to ensure proper cross-pollination. Most varieties will successfully pollinate other varieties. However, four—Bing, Emperor Francis, Lambert, and Napoleon, will not pollinate each other. Perhaps they're stubborn. Nevertheless, you can grow them. All it takes is another variety to ensure an effective fruit set.

Sour cherries, unlike their sweet cousins, are self-fruitful. Even one tree will bloom, pollinate itself, and bear well.

Perhaps one other drawback that has restricted cherry growing somewhat has been the worry that the birds will eat the fruits. That may be true with sour cherries, but not so much as you imagine with those sweetest ones. Birds, you see, prefer tart fruit. That's a fact of bird lore and life. Even when they sight your sour cherries beginning to ripen in the summer sun, you can avoid disaster. Convenient netting is available that fits neatly around the trees. In dark-green or black mesh, it blends into the foliage, fooling and thwarting those flying predators.

Cherries also have been known to suffer from diseases in the past. Thanks to plant breeders and their ability to produce virus-free stock, cherries can now be started better and free from one old-time problem. Improved pesticides and fungicides also make it possible to keep these trees thriving, well protected from other formerly troublesome diseases.

Two other factors should be considered as you mull over cherry growing as part of your outdoor plantscape. The tart (sour) type is fairly hardy, about as much so as most apple varieties. However, buds can be nipped in cold winters or by late frosts. No matter. A crop may be missed in some years, but trees will keep growing for future harvests. Sweet-cherry trees are slightly less tolerant of cold weather than peach trees. If you are living in a cold northern region, perhaps it is best to pass them by in your plans.

Three different rootstocks are used for producing cherry trees. *Mazzard* and *Mahaleb* seedling cherries and *Stockton Morello* softwood cuttings are those most frequently utilized by nurserymen. Mazzard is the main stock used on the West Coast for sweet cherries and others as well. Trees are larger and resistant to root-knot nematode, borers, and other problems.

Mazzard also is preferred for sweet and sour in eastern states where winter temperatures are not too extreme. However, in the East and northern areas as well, Mahaleb rootstock has proved hardier and more drought-resistant. The Morello stock is better suited for southern and western areas as a carrier for sweet cherries. It is semidwarfing to the tree.

Consider the characteristics and hardiness of each tree, and select the type you wish for its purpose in your landscape. For planning purposes, any land site that is suitable for peaches and grows them well, or an area that is suited for peaches in your state, should prove equally satisfactory for growing cherries.

Here are some leading varieties among *sweet cherries.*

Emperor Francis is a large cherry, red and dark. It resists cracking and has high quality. This is a promising new variety worth trying.

Hedelfingen ripens fairly late. It thrives in Canada and is a large, firm-fleshed black cherry also resistant to cracking in summer heat.

Ranier trees are vigorous and productive and bear early. They are hardy with large, firm, high-quality fruit. The skin is yellow with a pink blush, and sugar is equal to Bing cherries.

Napoleon is another hardy, high-quality yellow cherry worth trying.

Stella is the first self-fruitful sweet cherry. It has large, dark red, blackish-colored fruit, on productive trees, but is somewhat tender in areas with severe winters.

Van also has large black fruit on large, highly productive trees. The quality of these firm cherries is good.

Vogue also has large black, shiny fruit with a small pit. The flesh is firm and sweet. Trees are very productive but susceptible to brown rot.

Among tart or sour cherries, consider these.

Montmorency, Meteor, North Star, and Mesabi all can withstand fairly cold winters even as far north as New Hampshire.

Montmorency is a large tree, vigorous and upright but somewhat spreading, with drooping lower branches. The fruit is large, ranging from light to dark red. The flesh is pale yellow with a reddish tinge. It tends to be freestone.

Meteor is similar to Montmorency but has somewhat larger fruit of more oblong shape. Trees are medium-size, upright, very hardy, and productive. Nurseries provide a wider list suitable for various growing regions.

Early Richmond is a medium-size tree, vigorous and upright with dense round top somewhat spreading in habit. Fruit is large and dark red, with pale yellow flesh of good quality.

If peaches grow well on your land, cherries should enjoy the same conditions. They prefer well-drained soil and respond to fertilizing with nitrogen just as peaches do. Perhaps emphasis should be placed on draining. With cherries, possibly more than with most other fruit trees, it is essential to provide adequate draining. Soggy, heavier soils that hold water just aren't satisfactory. You can improve them, but it's an uphill fight to get them right for the best cherry growing. On the opposite side, sandy soils that dry out aren't right either. These are easier to improve with the addition of organic matter. But, sensitive as cherries are to late frosts, poor soils, and extreme winters, it pays to pick just the right spot.

Cherries enjoy full sun. It helps them brighten and sweeten to juicy perfection, whether tart or sweet. Sunny locations also dry off leaves faster, thereby thwarting mildews or other problems that may occur from time to time in warm, damp weather.

PLANTING-TIME TIPS

Sour cherries can be planted as close as 15 to 18 feet apart. Their spreading, graceful growth pattern fits them well to grace your lawn. Since they are self-pollinating, they can be used as specimens in various spots.

The larger, sweet-cherry varieties have a typical upright growth and require more growing room. Allow 24 to 30 feet between them. Do plan for several in a group, since they require a neighboring sweet variety for proper pollination. This combination of requirements may cause you to change plans and choose sour varieties instead.

Sweet-cherry trees usually grow large and upright. Some do have a slightly spreading habit as they reach maturity. One-year-old trees usually arrive as straight whips with few, if any, lateral branches. You can buy older trees, but it is easier to prune and train these youngsters. Also, there is little difference in the time they will begin to bear between one-year-old transplants and larger stock. However, local nurseries may offer balled and burlapped or container-grown sweet-cherry trees already through their first pruned shaping. Planted properly, they'll do well.

Plant sweet cherries as you'd plant apples or pears (see details in those chapters). Since cherries are somewhat more tender, be careful of the buds along the trunk. They can rub off easily if you aren't careful, thereby eliminating a chance for good laterals to form in the right spots.

After the tree's first year at your home, lateral branches should have begun to form. Retain four to five main ones spaced around the trunk beginning 18 to 24 inches from the ground. Choose only those that form wide angles with the trunks. Remove those that form narrow crotches.

Retain a central leader, and follow the general pruning directions

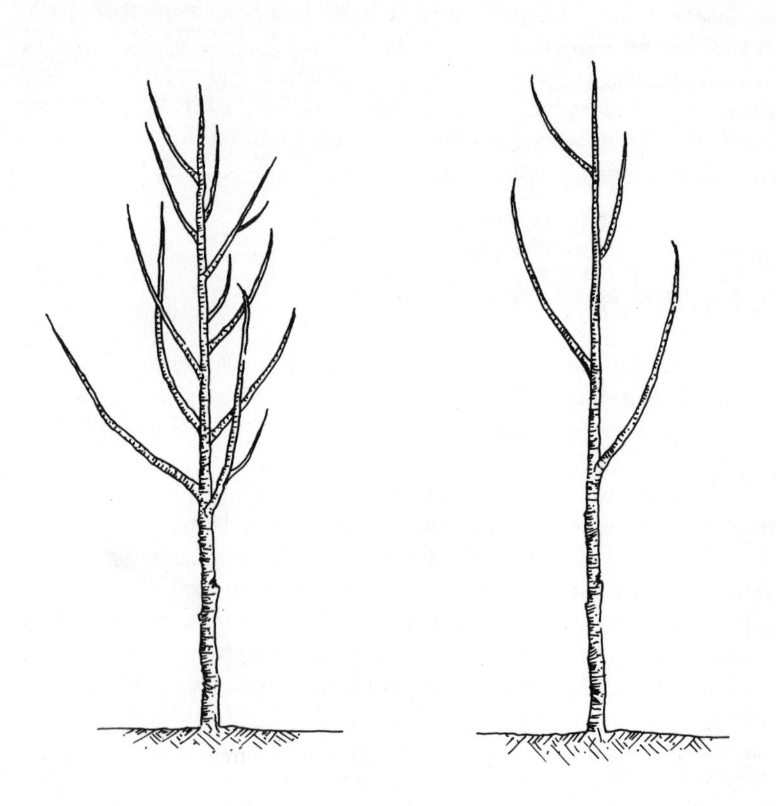

A cherry tree that has been growing in your yard for a year or so may look like the tree on the left. It should be pruned to resemble the one on the right.

you would for apple trees. The same basic steps apply each year in the training of sweet cherries. If you wish to try espalier, save that urge for the smaller, sour trees.

During the first few years, continue to select appropriate side branches until your tree has at least five well-distributed limbs. You may wish to do some thinning out to eliminate shoots that tend to grow up, cross, or turn to the center. Cut branches to form outward-growing limbs, eventually to achieve a more graceful and spreading appearance, even though these trees tend to grow upright. Once sweet cherries begin to bear, they'll require little pruning. Just remove damaged or diseased branches as you find them with clean cuts to prevent insect or disease entry into limb or trunk. Keep an eye on the overall shape and remove any suckers.

Sweet cherries form much less lateral growth than do sour cherries. That makes pruning easier. However, they tend to get tall. If you wish to keep them in bounds, head the leader and tallest branches back to about 15 to 18 feet high so they remain within a manageable 20-foot height. If long limbs produce a whorl of side limbs at long intervals, reduce these to two or three the first dormant season after they have formed. If you postpone this job for a few seasons, pruning these whorls from five to six at one point to a few later may cause some stunting of the tree.

SOUR-CHERRY PRUNING POINTS

Sour cherries tend naturally toward spreading growth. They should be pruned as you would a peach tree. Usually these young trees from a nursery are better branched than sweet ones. That's natural considering their future growth habit.

Remove weak, smaller side branches back to one bud. Keep the lowest branch no lower than 12 to 18 inches from the ground. Allow three or four side branches to grow around the main trunk. These will produce the basic scaffold for your mature tree.

Sour-cherry trees grow more slowly than peach trees do. That means less pruning is needed after the first few years of training. Prune sour-cherry trees lightly during these first five to seven years just to correct growth pattern and promote the tree's symmetry. Sour cherries may produce denser foliage and an overabundance of twigs. If that happens, do some thinning in order to give the tree sun and air.

As trees mature, extra attention is necessary to keep them in bounds. Reduce tall upright branches to form outward-growing ones. In all pruning of sour cherries, keep in mind the basic methods for peach-tree training. Reread the peach chapter as a guide. Your objective is to keep the top of sour-cherry trees open to allow lots of light to enter and provide good air circulation as well.

As a general rule of green thumb, young nonbearing sweet cherries should produce an annual shoot growth of 22 to 36 inches. Non-bearing sour-cherry trees will make annual growth of 12 to 24 inches. As these trees mature to bearing years, they should be producing 8 to 10 inches of new shoot length per year.

FEEDING CHERRIES

Tests by commercial growers and college researchers and reports from home gardeners all confirm the same fact: cherries, whether sweet or sour, respond to the same fertilizing program that promotes best growth and productivity in peaches. Rather than repeat that information here, I refer you to the nourishing of peach trees and recommend those directions. Sweet cherries need somewhat less nitrogen than sour varieties. That may seem odd, since they are larger trees, but it has been proved.

CONTROL PESTS PROPERLY

Cherries can be harmed by fungus diseases as well as by insects. Your first line of defense should be insect control, since insects can carry disease organisms from infected trees in the area to your trees.

Brown rot and leaf spot are most common. Mildews may occur during warm, wet weather. Modern fungicides in a preventative spraying plan can provide protective barriers against these problems.

Cherry aphids, plum curculio, and fruitflies are the more common insect pests of cherries. These too can be easily prevented with a careful pest-control plan. Details about the proper procedures are included in Chapter 18. In some areas, other specific problems may occur. It pays to contact local authorities from county agent to garden-center or nursery specialists. They know local conditions best and can guide you in your efforts to keep your cherries thriving.

Birds have an uncanny knack of sensing just when cherries begin to ripen. Netting to prevent this threat by birds is necessary, since cherries seem as appealing to them as they are to us.

Knowing when to pick cherries is a particular art. As they mature plump and ripe-looking, sample some. As soon as they seem sweet enough (or, in the case of tart ones, sweet enough to taste without wincing or puckering your mouth), begin to pick. Leave nets on the trees until your harvest is over.

Cherry leaf spot being the problem that it can be, it pays to give your trees extra attention when leaves fall. Rake the leaves and burn or dispose of them. That denies disease organisms and spores a handy spot to overwinter, on dead leaves beneath the trees.

Mice enjoy nibbling on cherry trees too. Clean cultivation several feet out from each tree is handy, although cherries will grow well on lawns with just a little mulch around the trunks to avoid bruising bark when you mow the grass.

Cherry tarts and pies and jam and jelly all can be yours from your own home-grown fruit. Cherries may take somewhat longer to become established, but once they're rooted, you'll enjoy them in tasty ways for many, many years.

9. Grapes Are Great

Wine may be fine when you dine, but grapes are great on any plate. And grapes fit into just about any landscape plan. You can grow them on fences, along walls, on a trellis or two, or in dramatic espalier shapes as distinctive living accents to your home grounds.

Because grapes are vines, they'll climb at will. By guiding them up posts, pillars, fences, and arbors, you can garden in the sky. Not only do grapes provide eye appeal but also you can just reach into your arbor and pluck a tasty bunch to tempt guests.

By selecting the proper varieties for your area, you can grow grapes in almost any part of our country. In fact, varieties developed by plant breeders from Gulf Coast states to California, from Great Lakes areas to Canada, provide a deliciously wide selection of types and varieties that will thrive in your garden wherever you live.

As with all plants, grapes have their own particular needs. When you fill these needs, grapes will reward you well with surprising abundance. They respond well to cultivation.

Whether you want grapes to eat fresh, to use for jelly or jam, or even to make your own wine, you'll find that they are a most versatile garden crop. As you plan your home vineyard, keep these points in mind.

Grapes prefer lots of sun, well-drained soil, and good air circulation. Their preference for climbing lets you save ground space by growing up. Adequate air circulation is more important with grapes than with other plants because it helps prevent certain diseases that can afflict grapes. Fortunately, plant breeders have developed fine new varieties that do have a certain built-in resistance to common problems. You also have the advantage of modern fungicides to keep grapevines healthy and productive.

Wine is becoming more popular in America. Wine lovers may argue the merits of the fine French wines compared to German Rhine wines.

Those in the know will verify that American wines made from luscious grapes grown in the vineyards here are just as robust and hearty or delicate and light as the finest imported wines. The reasons are clear.

One reason is that most U.S. grape varieties today are related to some of the most famous of Europe. Early settlers brought vines when they emigrated to the New World. Their knowledge enabled them to select ideal sites on gentle slopes bathed in beneficial sunlight, free from frost pockets. As they established the extensive grape and wine industry of America, they also improved on the old-time varieties. Plant breeders crossed the best of available types. They bred in hardiness, flavor, productivity.

There are so many good varieties of grapes of white, red, blue, and green types for table use, jelly, jam, preserves, and wine making that it would require pages to list them all. In this book you will find lists of leading and reliable nurseries and mail-order firms. Often you can buy excellent grape varieties locally that do well in your area. If not, you may wish to get catalogs from mail-order firms.

I will list some of the better varieties as recommended by our good gardening friends for various regions. These have been tested and found to be quite satisfactory, some even exceptional, especially for home gardens. With all things, there are good growing years and poorer ones. So it is with grapes, depending on the combination of sun, moisture, plant food, and such in a given year.

Be aware that plant breeders continue to develop and introduce new, improved grape varieties. Some are designed primarily for large-scale commercial growing, ease of harvest, and wine making. Others are more suited to home-garden cultivation. The top mail-order nurseries have built their reputations on providing quality fruit trees, grapevines, and berry bushes to customers over the years. As you shop, consider the purpose of your grapes, whether for table use, jelly, or wine making.

PLAN YOUR GRAPESCAPING

Grapes, being vines, offer unique qualities for your garden plantscaping. They'll grow up not only trellises, fences, and poles but even tall trees if you don't prune them regularly. That's their natural growing habit. Take advantage of it to the fullest.

You can also train grapevines into beautiful and dramatic designs.

This technique of growing living sculpture is discussed in more detail in Chapter 16. In this chapter, we'll concentrate more on the special uses and places for grapevines in your living landscape.

When you consider grapes, remember they are sturdy, stubborn, and amazingly long-lived plants. Some vineyards have been producing for decades. In Europe, there are vineyards that trace their heritage back centuries. Keep firmly in mind the longevity of grapevines as you plan their permanent places in your fruitful landscape.

Grapes can withstand drought and cold. They can succeed even in quite rocky, seemingly infertile soil. Given reasonable care, grapes often outlive those who plant them, and continue rewarding the generations that follow.

Remember, please, that your plants will respond best when you provide them with the conditions they require and the environment they prefer.

The two most important considerations for grapes are good, full sun and adequate air circulation. They like well-drained soil and growing room free from frost pockets. Avoid areas where icy air settles to nip buds in spring and plants in fall.

Sun and air circulation should come first. You can usually improve drainage and soil, add fertilizer and water, but it's difficult to cut down trees or move a garage to give grapes the sun and air they must have.

Probably the best and most common use of grapes is along an existing border fence or on a special trellis. Grapes are perfect for a property border or boundary marker. They add privacy, screen undesirable views, and add eye appeal as well as taste appeal when they bear their plump clusters of fruit.

Grapes aren't fussy about the type of fence they use for their support. But support they do need. They'll look fine along a post-and-rail or a woven-wire fence. It's best to avoid small-mesh wire like chain-link fences. Tendrils intertwine with the wire, and you just can't get pruning done properly then.

Avoid a picket or stockade fence, since both need periodic painting and restrict air flow around your vines. True, you can tie arms of the vines to these fences and untie them in late fall or early spring before leaves form to do the fence painting. But why ask for more work?

A wire trellis works well for grapes. Using wood posts and #10 or #12 fencing wire, you can create a living fence of vines that hides posts and wire supports quite well. Farm fences also support grapes handily.

The fan system has been popular for home grape growing too. You can train vines to walls and fences with this hand-shaped growing method. It's easy to make a trellis of poles too and train the vines along each support, just as you would a climbing rose bush. Strong wires stretching in fan or even wagon-wheel shape from a T-bar support create an artistic effect. Grapes will respond to more sculptured designing than most fruit plants. Other techniques are shown in Chapter 16.

Many people prefer a more decorative approach combined with practical use. Arbors are the answer. You can build arbors of all shapes and sizes. "Sitting under" size we find the best. All you need is sold 4 × 4 timbers topped by 2 × 4 or 2 × 6 lumber. On that basic frame you can add 2 × 2s to complete a lattice work. Or use overhead and side wires instead of the 2 × 2s. Vines eventually will grow up, spread out, and shade your outdoor sitting area. Come summer, you can do as the ancient Romans did: reach up or out, right or left, to pluck a tasty grape or two. Design the arbor to match your home architecture.

Since grapes are so long-lived, it pays to erect a sturdy permanent arbor, fence, or trellis. An arbor is ideal when you have adequate room. In fact, the tops can bear the grapes. Along the sides you can plant currants or other berry bushes if you like.

Logically, the next step in planning your grapescaping would be how to plant. However, since pruning is perhaps the most important key to grape-growing success, planting tips will come later.

The four-cane Kniffin system is perhaps the favorite of most grape growers and probably the easiest to use for home gardens. It is based on an ordinary two-wire fence. Wooden end posts should be at least 6 inches in diameter. You can use metal posts, but wood looks more natural. Select sturdy end posts 7 to 8 feet high. Set them 3 feet deep and brace them well. These can be along a property line, as a divider of a play area from another garden spot, or along a path or drive.

Select cedar, redwood, or cypress. They resist rot. If you use other types, be sure they are treated to prevent rotting in the ground. End posts should be cased in concrete or set deeply enough to keep wires taut. Posts between the anchoring end posts should be set 2 feet deep. That seems extra deep, but heavy crops can pull out shallow posts.

Space posts 8 feet apart so you can plant two grapevines between each set of posts. Trellis wire should be smooth and galvanized to withstand weathering. Use #10 or #12. Once posts are set and end posts braced and spaced properly, connect your wires. The bottom wire should be 30 inches from the ground, the top one 30 inches above the first.

If you wish to screen an unwanted scene, you can add another wire

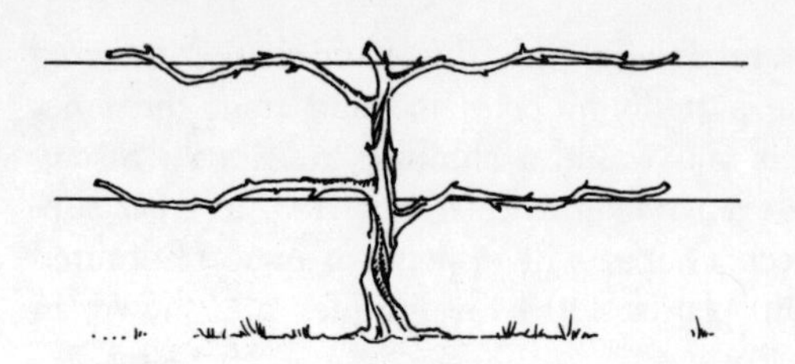

FOUR ARM, SINGLE TRUNK KNIFFIN SYSTEM.

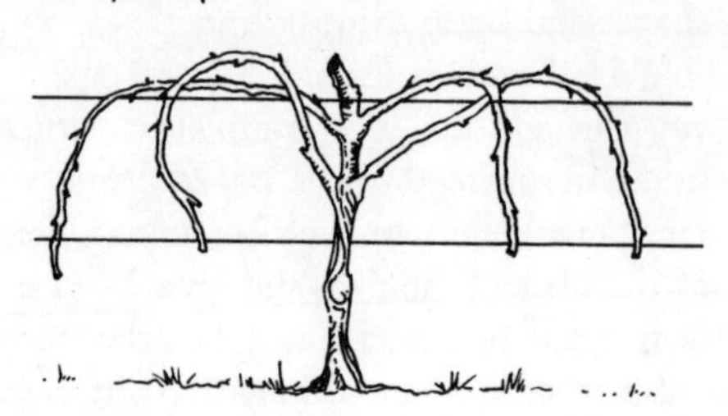

FOUR ARM, SINGLE TRUNK UMBRELLA SYSTEM.

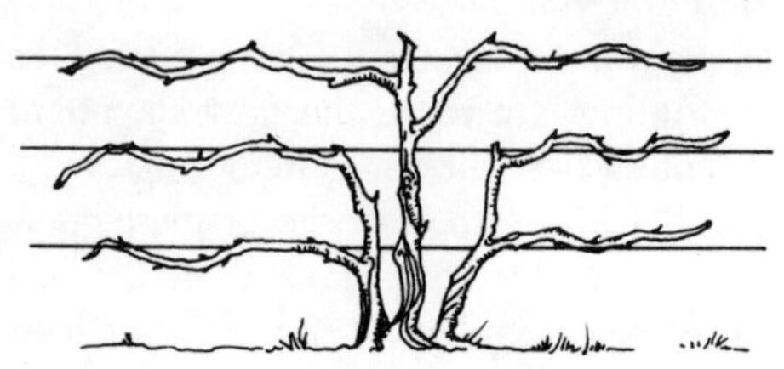

SIX ARM, THREE TRUNK, THREE WIRE, MODIFIED KNIFFIN SYSTEM.

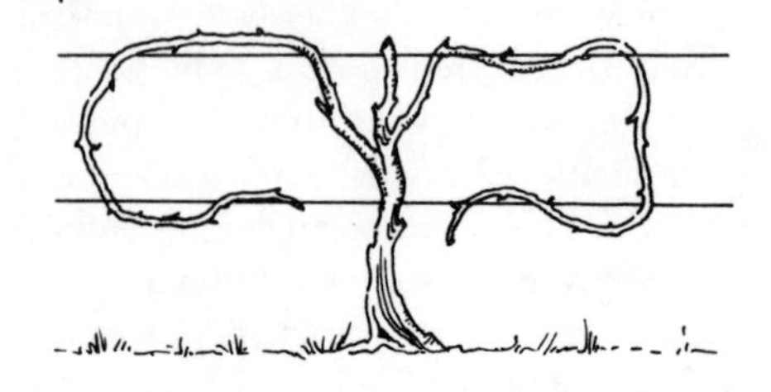

TWO ARM, SINGLE TRUNK UMBRELLA SYSTEM.

You should decide in which of the several common forms you will prune and train your grapevine, long before it reaches this size.

30 inches above the second, making a so-called six-arm, three-wire modified Kniffin system.

Examine the diagrams in this chapter carefully. You can adopt a four-arm or six-arm system or choose a modified-umbrella system as you prune and shape your grapevines. The same basic pruning is required, whichever system you select.

Knowledge of grape-growing terms can be helpful. The trunk is the main permanent stem of the plant. Arms are the short side extensions of the main stem. Shoots are the immature soft stem growth of the current growing season. Shoots arise from buds on wood, one or more years old, and bear the leaves, the flowers, and the fruit. Canes are the mature shoots. They become woody after growth has ceased. Fruiting canes are the one-year-old canes that are capable of and suitable for bearing fruit. Spurs, on the other hand, are one-year-old canes, preferably originating near the trunk, which are shortened in pruning to two buds. From the spur buds, shoots develop to become canes. You select one of these as a fruiting cane for the following season. In this way you renew the fruiting wood. Suckers are the undesirables of a grape plant. They usually arise from the lower part of the trunk and always should be removed.

Now take out your pruning shears and keep this book in hand. If you have grapevines planted, follow these next steps carefully. If you're planning your mini-vineyard, refer to these tips in the future.

A grapevine that's overgrown, like this one on the left, should be brought under control by proper pruning. At right: same vine pruned to six-arm form.

Examine your plants. The first year, you should select the best cane and shorten it to two buds. Remove all other canes along that wire. Sometimes a particular plant is especially vigorous and has one cane reaching the top wire and beyond. If so, cut it at the height of the top wire and tie the cane to it. Grapes may seem slow to get a roothold and get growing, but have patience. Proper pruning will reward you for many, many years ahead.

The second year, select the two most vigorous canes along the top wire and the bottom one. Remove all others. Tie these canes to the wires.

When it is time to tie the canes to the supports, whether wood stakes, wire fences, or metal posts, use strong string or coated wire. Winds do blow, and wire ties can cut and damage the canes. Cord or plastic-coated ties protect the tender skin and bark.

During the second growing season you can select the best, strongest, most vigorous canes for the future. When shoots are several inches long, remove all but three or four arising from near each wire. Also remove all those on the trunk that are not near the wires. You may feel that this is being heavy-handed with the pruning shears. Not so. With grapes, more attention to pruning is necessary to encourage maximum performance of the few canes that are left. Surprisingly, this regimen stimulates the plants to be more prolific in their bearing habits. After all, that's what you want, a real abundance of grapes, not necessarily lots of foliage with little fruit.

By the third and fourth year, your grapevines should be taking their desired shape. In the dormant period preceding the third year, your plants should consist of a main trunk with several canes along each wire. Select two of the best canes at each wire and remove the others. Shorten the selected canes, leaving four to five buds on each. These buds left on the short canes will produce the shoots which will bear some fruit. They also will be the source from which you'll choose next year's fruiting canes. Grapes, like other berries, sprout new shoots and bear best as these become fruiting canes the second season. That means you must prune each year to keep grapes most prolific.

In future seasons, you should have two or more canes extending in both directions on each wire. Select a renewal spur at each of the four arms, from canes nearest the trunk. This is important for maintaining fruiting wood close to the trunk and keeping the entire plant within its allotted space. Grapevines will ramble far afield and be less fruitful if not kept in check by annual pruning.

Each year onward, select the fruiting canes of moderate vigor, diameter, and length, originated near the trunk and reasonably close to the appropriate trellis wires. Remove all other growth, including sucker growth arising on the lower trunk. Shorten selected fruiting canes to six to ten buds. Leave more or fewer buds depending on plant vigor, up to twelve on strong vines.

In future years, pruning is similar. Your objective is to remove old wood so that the second-year fruiting canes can be most productive. Pruning stimulates productivity. It may hurt your feelings to cut away those lovely old branches, but it benefits your grapevine. You can increase the number of buds left per cane each year, up to twelve or fifteen once your plant is well settled and growing vigorously. By fertilizing properly, you'll be able to keep the vine in good health and as productive as it can be.

For the umbrella system, begin the same as with the Kniffin method. However, the objective is to select the tallest, most vigorous canes that reach the top wire. They are then allowed to flow down to the lower wire. Each year, you repeat the pruning to allow the most vigorous fruiting canes to flow so they can be fastened to each wire as shown in the sketch. You must, of course, let some new shoots form so they become the second year's bearing wood.

Pruning for the fan system or an arbor follows the same idea. You should select the most vigorous shoots and canes, tie them to the appropriate supports, and remove all other side shoots. By removing these extra shoots and canes you force the growing strength of the plant into the remaining ones you selected as bearing wood. That living strength is what produces the abundance of blooms and fruit which, of course, is your objective in this fruitful landscape plan.

PLANTING GRAPES

Once established, grapevines seem to last forever. But they do try your patience in their initial slowness to get growing. It pays to plant them right to give them the proper start and encourage a strong, sturdy roothold for the years to come.

Well-drained soil is important. If your selected area is sodded, till or spade it deep. You can improve soil by adding compost and well-rotted manure mixed with the existing topsoil. If soil is heavy, be sure to mix sand, peat moss, and compost with it to provide the drainage your vines need. Prepare your planting spot in fall. We have found that placing organic matter on the area to be planted and deep-tilling with our Troy Bilt rototiller pulverizes the ground well, incorporating organic material into it. This rear-mounted tiller is easy to handle and can be controlled well, either in row tilling or to dig individual spots for tree and shrub planting.

If your soil is short of lime, apply the amount needed as indicated by a soil test before tilling. You can also spread fertilizer or manure on the surface and till or dig it under the previous fall. This procedure would help digest the organic matter, providing a fertile foundation for spring planting.

Select vigorous one-year-old plants. Nurseries usually list these as one-year #1s. Two-year-old plants are fine, but cost more. They will, however, give you a head start toward harvest time.

SOME GOOD VARIETIES

In this book, when I use the term *table use,* I am referring to fruit for eating fresh as well as for jelly, jam, and juice.

In Southern areas, here are some fine varieties of bunch-type grapes to consider.

Blue-Black types include *Van Buren,* with hardy, vigorous vines for early table use. *Buffalo* has excellent quality for early table or wine use. *Freonia* is mid to early season with hardy, vigorous vines as a Concord-type grape. *Steuben* is midseason for table or wine. *Concord* is mid to late season and has a universal standard of quality for many purposes. It is the most widely grown in America.

For *White* types, try *Himrod* or *Interlaken* or *Seneca.* All are early and of high dessert quality. Himrod and Interlaken are seedless; the latter is restricted to southern areas. Seneca is somewhat difficult to grow, but one of the best table grapes.

Niagara is midseason with large fruit for table or wine.

Of *Red* types, *Delaware* has high sugar content and matures in midseason. But the vines lack vigor. *Catawba* is late season but has vigorous vines and large berries, tasty for table or wine uses.

Many nurseries, among the best being Stark Brothers in Missouri, can provide fine French-American hybrids. Their catalog advises which varieties of all types are best for various parts of the country.

Among muscadine grapes for Southern to more mid-America regions, varieties worth trying are *Carlos, Magnolia, Noble, Fry,* and

Hunt. As you shop for appropriate varieties of muscadine grapes, be aware that those listed as female vines will not have fruits if they are planted alone.

For more northern areas, similar to the Great Lakes region, white Niagara with its large, compact bunches of berries, red Delaware with its fine table and wine uses, blue Concord, that popular favorite, and red Catawba with medium-size berries are good choices.

Here's a recommended list of grape varieties with their relative cold-hardiness. It is provided by C. A. Lunger, a good growing horticulturist from New Hampshire, where grapes are making a comeback in home gardens and even commercially.

MOST HARDY

Beta
Blue Jay
Red Amber
Clinton
Brighton
Concord
Fredonia
Worden
Seibel 1000

HARDY

Niagara
Delaware
Van Buren
Buffalo
Foch
Agawan
Other Hybrids

MEDIUM HARDY

Baco 1
Portland
Steuben
Other Hybrids
Golden Muscat

LOW IN HARDINESS

White Riesling
Interlaken
Romulus

You can, of course, decide to experiment with the more exotic varieties. Among these are *Rieslings* and *Chardonnay, Muscats* and *Beaujolais.* Which varieties you grow depends on your purpose. The list is long, so take your time, get advice locally on those that prove best in your area, and plant away.

Rows should be 8 to 10 feet apart if you have room, with vines 4 to 6 feet apart in rows. In commercial vineyards, plants may be farther apart. For home grounds, considering that you will be giving your vines tender loving care and lots of nourishment, you can plant grapes

A grape plant should be planted in a hole of proper size that's then filled with improved-soil mix. Prune as shown. As the vine grows, it should be tied to support.

closer together. Before planting, you can improve soil as you would for berries and tree fruits.

For proper planting, open a hole in the ground large enough to accommodate the roots of the grapevine as they spread naturally. Usually this will be 15 to 18 inches across. Plant vines the same depth as they grow in the nursery. You usually can see this point on the trunk. Keep the graft point of grafted grapes *above* ground to discourage suckers of an undesirable variety from sprouting and taking over.

Add soil and firm around the roots. Tamp it down. It helps to shake the plant gently to settle soil around roots. Then tamp again, and add water and more soil. Finally, fill the remaining space and leave a saucer-shaped basin around the newly set plant. This lets water collect to help the vine get started right.

Water your newly planted vines regularly, especially in dry periods, until they are well started. Then be sure they receive an inch of water each week, especially at fruiting time to encourage the plumpest, sweetest grapes.

CULTIVATION AND CARE

Moderate amounts of fertilizer stimulate young vines to grow and maintain vigor as bearing plants. There are, of course, wide variations among soils. One fertilizer program will not be suited to all areas. You can, however, adjust these general rules successfully. The first year after planting and before growth starts, apply ½ cup (about a quarter-pound) of 10-10-10 in an 18-inch circle around each vine. Repeat this amount monthly until mid-July. You can also add compost or rotted manure. This helps mulch away weeds and will improve the soil, bit by bit, as you cultivate it into the ground at the end of each season.

The second year, double the first-year amounts. Apply 1 cup of 10-10-10 in an 18-to-24-inch circle around each vine, and repeat monthly till midseason. Either clean-cultivate around vines to remove competing weeds or mulch well.

For bearing vines, begin to apply fertilizer in early spring. Spread 2 to 4 pounds of 10-10-10 around the plants and the area beneath the vines. Roots will be spreading and feeding underground over a wider area to support mature plants. Repeat with 2 pounds per vine after the fruit sets.

Grapevines do have somewhat shallow roots as well as deeper ones. If you cultivate rather than mulch, be sure to keep cultivation shallow to avoid damaging roots close to the surface.

Grapevines will tell you when they are well fed. Vigorous growth and nicely plump buds indicate that they are happy. Fruiting canes ¼ to nearly ½ inch in diameter, slightly larger than a pencil, and 5 to 8 feet long on most varieties are desirable. Gear your fertilizing program accordingly.

WATCH THOSE PESTS

Several pests and diseases can damage your grapes. Usually it is necessary to apply some sprays during the season to grow fruit of consistent good quality. The grape berry moth is a common enemy. Small greenish larvae feed on the berries. Leafhoppers, those small wedge-shaped jumping insects, feed on grape foliage. Mealybugs and flea beetles also may attack grapes. In eastern states where it exists, the Japanese beetle can be a serious problem at times, skeletonizing leaves. All these pests can be thwarted with *proper* use of pesticides.

Black rot is a fungus disease that attacks foliage and often fruit as well. Berries become blackened and shriveled. Downy and powdery mildews also may damage grapes. All can be avoided with a proper preventative program of pesticide application. Some problems are more common in one area than in others. Your county agricultural agent and garden-supply dealer can tell you which pests are most common where you live and which multipurpose pesticides will provide the protection your grapes need.

Mildew and fungus problems often can be prevented with proper air circulation and avoidance of watering vine foliage, especially in warm weather. To irrigate grapevines, it is better to use soaker hoses on the ground than sprinklers.

Despite what you may think, grapes don't require direct sunlight on the berries for development of the best color and sweetest taste. Color development is governed by the amount of sugar produced by leaves and translocated to the fruit. Don't worry if all your bunches of grapes are completely hidden from the sun. It's the leaves that vitally need sunlight to produce the sugars.

Birds often enjoy grapes. They seem to favor some varieties more than others. Since grapes themselves don't need sun to mature well to their flavorful peak, you can surround growing bunches with brown paper bags when clusters are half grown. Tie the bags securely with string or twist ties to thwart birds.

A new type of netting, made of nylon and plastic, is colored green or black. It blends with the foliage and also prevents your feathered friends from eating into your grape crop.

HARVEST TIPS

Color isn't the only indicator of grape maturity. Actually it is less reliable than other evidence. When grapes are ripe, seeds usually change from green to brown. The cluster stem will turn slightly brownish and wrinkled. That's when berries reach their peak sweetness. Some varieties may shatter from the bunch before fully ripe, so keep an eye on them. Excess rain may cause split skins. Taste a few grapes periodically when they approach maturity. Pick when they please your taste buds.

PROPAGATING FAVORITE VARIETIES

Although you can buy a wide range of colors, types, and varieties of grapes, their ease of propagation makes it possible to create your own new plants. Perhaps friends and neighbors have superb varieties. Ask them for some cuttings. After all, they'll be pruning away lots of canes each year. Or ask them if they can do some simple layering to provide you with new plants.

To propagate new grapevines from cuttings, take cuttings during the dormant season. It is best done in early winter or early spring. Select sturdy canes and cut them in 10-inch lengths, with three to four buds. Slice the bottom, closest to the trunk, at a slant. Cut the top square. Tie cuttings into bundles. Place them vertically in a trench with the bottom ends several inches in the soil. Keep the soil moist, and mulch with straw, leaves, or grass clippings to provide winter protection. By the time the ground is ready for planting in spring, cuttings should be calloused over on the bottom and ready to root quickly. Place them 4 to 6 inches apart in a loamy soil that holds moisture fairly well. Insert cuttings with the top bud just above the soil surface, the other buds below the soil. In about one year the cuttings will be ready for transplanting.

For layering of your own vines or those of a friend, dig a shallow trench near the vine. Bend one or two canes from each vine into the trench with the tips sticking up. Cover with soil. Within a year, roots will form from underground buds. Then, simply sever the canes from the parent plant and you'll have new stock to plant around your home grounds.

Grape growing is fun. As you develop your skills, you may branch out into grafting of new and better varieties on more common root-stock. You may try your hand at wine making and become a connoisseur of this age-old art.

Grapescaping can add new dimensions to your outdoor-living areas. When you enjoy the fruits of the grape for the table, in jam and jelly or otherwise, you'll most likely find new areas to plant.

10. Really Ripe Raspberries

When you give friends raspberries right from your garden, they'll marvel at your growing ability. Home-raised raspberries are among the sweetest treats a garden can yield. You have a choice of red, purple, black, or amber-colored juicy fruits from these bushes. Each is distinctively different and delicious.

Raspberries have versatility. They thrive in almost any type soil except the lightest sands and heaviest clays. Even those soils can be improved by following the steps in Chapter 2. That means that raspberries can be a part of your home plantscape plans using those areas that are less desirable for other trees or crops. Raspberries, however, do best in areas from horticultural zone 4 into the northern states. That's from the middle south to the Canadian border.

You have a choice of shapes and forms too. Black raspberries and purple types have a distinctive spreading and drooping habit. They can be used in small groups as a mound effect in the landscape. Even a few yield well to provide several quarts of good eating every year.

The red raspberries, normally larger and more succulent, are better used along a property line or in hedge effects. These varieties—and the choice among good ones is wide—tend to grow upright in thickets. You can train them somewhat, but they'll do just as well with little care in an unused part of your land. If you have room, try several double rows along a walk or path to your vegetable patch. Another good use is to produce a screen effect.

Raspberries are somewhat like blackberries. They have a fondness for wandering by sucker growth arising from their roots, which is typical of the red varieties, or by propagating themselves from tips as black and purple canes touch the ground to set new plants from growing tips. It's easy to keep them in bounds, but you must pay attention to pruning once or twice a year.

SIMPLE SITE SELECTION

Although raspberries can grow reasonably well on less than the best soils, they'll reward you more abundantly (as all crops will) if you pick the most desirable site for them. A well-drained soil with liberal amounts of humus added year by year will increase their productivity. Sandier soils, including sandy loams, tend to dry. Unless you provide ample moisture at fruiting time, raspberries just won't be as juicily plump and sweet.

Good sun is important with red varieties. The black ones can do quite well with partial shade provided they get a fair share of sunlight for several hours each day. In our raspberry patches, black varieties along the edge of a pine woods and some near a dogwood planting produce quite well. Another group along a mini apple and pear orchard received good sun half of each day when trees were in leaf and still performed deliciously.

A site on ground higher than surrounding areas is desirable for raspberries. Cool spring air drains away from such an area, so that nipped buds and reduced crops are avoided.

Once you have established raspberries, you can improve the soil and fertilize and water the bushes well each year to keep improving their performance. In comparing level sites to gentle slopes, many gardeners believe a slope somehow does improve success with these berry bushes.

Before planting, prepare the soil well. It pays to pick a site on which cultivated crops were grown the year or years before. That way, soil is usually in better condition. If you must choose a site in sod, dig or till the sod under the fall before the spring you wish to plant. But first spread lime and 10-6-4 or similar-analysis fertilizer on the area. When you turn the sod under, this fertilizer helps decompose the organic matter, including old plant roots. With raspberries as with most other berry crops, the more you can improve the aeration and structure of the soil with increased amounts of organic matter, the better the plant growth will be.

If you can prepare the berry site in early fall, try to plant a cover crop of rye grass. Its roots will penetrate deep and will open soil well. Then, merely till this "green manure" under come spring planting time. If animal manure is available, it too can be plowed under to incorporate even more organic material into the soil.

You can mulch around established raspberry plantings, but clean cultivation several times a season seems to produce more favorable results. So it pays to incorporate as much organic matter as possible into the chosen planting area before planting. Once established well, raspberries will renew themselves for many years as a permanent part of your garden.

Raspberries are sold usually as rooted canes. Once you have some mature plants, they'll provide ample new stock for future plantings. However, in the first planting, be certain to insist on certified disease-free stock. Raspberries are susceptible to crown gall disease, anthracnose, and some other wilt diseases. Be sure the plants you buy are from a reliable nursery and certified virus-free as well. Discard any canes that have wartlike growths or galls on roots or crown areas. It also pays, as with blackberries, to destroy any wild raspberries in the area. They may tolerate and harbor raspberry diseases, acting as a carrier to infect your domesticated varieties.

SPACING FACTORS

In rows, plant your raspberries 3 to 4 feet apart. Space rows 6 to 8 feet apart for red varieties.

Black and purple also respond to hill cultivation. This means simply planting in such a way that a bushy mound results and can be maintained in that desired shape. If you elect to use the hill system with these types, plant young rootstock 5 feet apart each way or allow a 5-foot diameter for each plant to mature well. If you promise yourself to prune raspberries regularly, you'll be rewarded in two ways: (1) they'll bear more abundantly as you stimulate their desire to grow back, and (2) they'll remain tidy rather than grow into tangled thickets.

You can also plant red, black, and purple varieties closer together to encourage faster fill-in of the rows or corner in which they are to become established. Reds can be planted a foot apart and in hills 3 feet apart. After that, attention to pruning will keep them in bounds.

SELECT BEST STOCK AVAILABLE

Research by Dr. R. H. Converse of the U.S. Department of Agriculture at the Research Station in Beltsville, Maryland, has revealed that raspberry plants may be infected with viruses. These plant diseases may not produce symptoms. However, they may reduce plant vigor and yield. The USDA researchers find that leaf grafting of one variety to a sensitive indicator plant will reveal if the virus is present. By extensive testing, plant scientists have found some that are free of all known viruses. Heat treatments have eliminated viruses also.

When you buy your foundation stock, be certain that the nursery guarantees that it is from virus-free parent plants and is certified disease-free. That precaution saves future worries and ensures hardier, more abundant plants that will yield most rewardingly.

Among red varieties, *Heritage* is a vigorous grower and everbearing. It produces many suckers to fill in rows quickly. Plants are winter-hardy and produce moderate summer crops. Then you get a bonus, a fall crop of medium-size, very firm, excellent-quality berries. In areas as northerly as New York State, berries may yield into September, until a hard frost occurs. Canes are sturdy and erect and seldom require support.

Hilton is the largest of all red raspberries. It maintains size throughout the season. Berries are long, conical, medium red, firm and of good quality. It ripens in late midseason and is vigorous and self-supporting.

Newburg has been around for some time. It produces firm berries well with fine quality and good flavor.

Taylor is another excellent variety for home use. Plants are tall, vigorous, hardy, and productive. Berries are large, conical, firm, red, and tasty.

You can also select *Latham, Sunrise,* and *Southland,* among red types. *Citadel* is a midseason producer of large berries and grows vigorously. *Comet* is a promising hardy variety, somewhat resistant to anthracnose and spur blight. *Milton* is a late variety, more tolerant to mosaic than most with tall cane growth, but not as hardy in colder areas as others.

Madawaska Red is fine too. It has a tough cane and is hardy in northern states into Canada. Berries are huge and ripen in midseason. *New Hampshire* is another midseason ripener with well-flavored fruit of large size.

Boyne is a new red variety from Canada. It takes cold well on hardy, strong canes. Fruit is flavorful.

August Red is an everbearing type for northern areas. It begins a crop in late July and continues for weeks to bear plump, delicious red berries.

BLACK RASPBERRY VARIETIES

You have a fair choice of these glistening, sweet delights for home growing too.

Allen is a newly named variety that is vigorous and productive. It bears large and attractive fruit with a large portion pickable at one time.

Bristol is well known in northern areas for its large, firm, glossy, and high-quality berries. Plants are hardy, vigorous, and productive.

Dundee has become another principal variety for commercial growing. It does well in home gardens, producing vigorous plants even on imperfectly drained soils, and is similar to Bristol.

Huron has large, glossy, attractive berries on vigorous, hardy plants. It is not too susceptible to anthracnose.

Jewel is a relatively new variety dating from 1973. It is productive and not susceptible to any serious disease and only slightly affected by mildew. Fruits are large, glossy, and high-quality. Plants are vigorous, hardy, and nicely productive.

Logan or *New Logan* is early to ripen and yields heavily.

Black Hawk is very late but productive and produces fruit for fresh use, freezing, jams, or jellies.

PURPLE VARIETIES

Yes, some raspberries are purple. They're deliciously different. Try some if you have room, although red and black deserve first-place votes for their vigorous bearing and productivity.

Clyde is a medium-to-late-ripening, moderately hardy, and consistently productive one. Berries are darker than medium purple, firm, and tart.

New York 905 is a promising new purple raspberry in its introductory stages. It is not a true purple, but resembles this type more than red or black. It has large, round, firm reddish purple berries. Stocks are scarce at present. The Geneva Fruit Testing Association has them, and they're worth trying.

Sodus is another purple variety, very productive on hardy, drought-resistant bushes. It is good for dessert, freezing, or cooking but is somewhat susceptible to mosaic and verticillium problems.

If you like light-colored raspberries, *Amber* is a very vigorous variety with large, amber-colored, sweet-tasting berries.

September also is light, but brighter red in color. It bears one crop early and another in early fall, except in northern areas where early frosts may hit the plants and damage berries before the major fall crop is fully ready.

PLANTING POINTERS

Avoid using old stocks from neighbors' plantings, which may bring along disease problems.

Raspberries are self-fruitful. Once you set them well, they'll bear year after tasty year. Several varieties offer different tastes and also improved cross pollination. Selecting several also provides the pleasure of early-, mid-, and late-season harvests to give yourself these bountiful berries over a prolonged period.

One-year-old rooted stock is the best buy. These are young suckers from the current season's growth. Set them anytime in the spring. Since raspberries are softer-caned than blackberries, it helps to keep roots moist and shape a ball of soil around the small roots when planting. You can spread the roots too, but always keep them moist. Fill the soil around the new plants, tamp it firmly to settle it, and water well. In transplanting suckers from your own established area to a new one, dig and plant suckers immediately. Drying out will make them suffer and you may lose a large proportion if they dry.

After planting, prune canes back to half their size. Keep watering them every few days until new buds sprout on the canes. After that, once a week is sufficient as they establish a sturdy roothold. Mulching in a circle around new plants the first year, about 18 to 24 inches, is advisable. However, raspberries do seem to respond best to clean cultivation after they are well-rooted and growing vigorously.

All raspberries are shallow-rooted. When you cultivate, be certain to do it lightly. Just remove weeds by scratching the soil to avoid damaging the feeding roots barely beneath the surface. Sometimes just hand-weeding is best.

In all the years we've raised raspberries, we have tried both clean-tilling and mulching. Being something of a mulch enthusiast for all its values, I still prefer to use a light layer of compost or peat-moss mulch on raspberries. They bear so well when fertilized properly and pruned yearly that I prefer to sacrifice some yield for the labor saving of mulching. All raspberries share a unique habit when you cultivate too deeply. You may cut roots, thereby encouraging many new suckers to grow. This characteristic is more common with red varieties than with others. Soon you have a tangle rather than the tidy and productive planting you want. For this reason alone, mulching has its advantages.

Perhaps a word here about suckers is in order. When that term is used with berries like raspberries and blackberries, it is a favorable term. It means those new canes that sprout to produce next year's fruiting wood. They are also useful for transplanting, thereby increasing your berry potential.

However, when the term *sucker* is used in referring to fruit trees, it means something quite different. I wish things weren't that way, because it makes gardening a bit confusing. But that's how gardening language has been for years. With fruit trees, suckers are those notorious, unwanted shoots that sprout from branches and limbs and even trunks. On berry bushes, on the other hand, you need an annual supply of suckers sprouting from the underground roots to ensure a consistent stand of canes to keep the planting productive year after year.

FERTILIZING RASPBERRIES

Although they can produce on poorer soils, raspberries are another fruit that responds well to tender loving care. That care includes providing manure or commercial fertilizer along their rows or around the plants to nourish them. It also includes adding compost as a mulch and lightly raking it into the soil each fall to improve soil texture and structure as well as to provide minute amounts of nutrients.

If you can get manure, that's grand. Use it, about 2 inches deep along the rows. You might introduce some weed seeds along with the barnyard manure, so be alert for weeds. Pull them out gently whenever they sprout.

Raspberries can use a moderate amount of fertilizer the first year. Here's a guide. For every bush, spread ¼ cup of 10–10–10, or less of 16–16–16, in a circle around each plant. You can apply this also as a band on each side of the row. The second year, when new suckers have sprouted and your raspberries are setting their permanent pattern, apply 3 to 5 pounds of complete 10–10–10 along each 100 feet of row. Split the amount so each side gets half the total. On black and purple raspberries grown as hills or bushes, a cup or two of this fertilizer around the bush is sufficient in spring.

Avoid late fertilizing in summer. That only increases tender cane growth and contributes to winter kill. It may also encourage excess sucker growth, which means more pruning work for you.

The first two years, your raspberries may seem to be off to a surprisingly slow start. Take heart. That's their way. Use that space between rows or around bushes for other plants during this period. If you don't wish to grow other crops, do plant a green manure crop of rye, oats, clover, or other cover crop. Then dig these green manures under in the spring. That way, you'll have been improving nearby soil areas so that as your raspberry plantings do expand the third and future years, they'll find excellent soil conditions to welcome their expanding permanent root systems.

If you mulch your raspberries with sawdust and wood chips or similar materials, remember that these cellulose mulches need extra nitrogen to decompose properly. Spread a cup of nitrogen fertilizer on top of the mulch around each plant or along every few feet of row. It will do wonders for the plants as it helps the woodier mulches perform their services.

PRUNING POINTERS

The first year, cut red raspberries to within 6 to 8 inches of the ground. Your plants will become rooted faster for long-term benefits if you hand-pick any blossoms that occur the first year. The second year a partial crop will form. From the third year on, you'll be rewarded by buckets of berries.

During the second year, remove any broken and damaged canes. If some have grown too tall, you may top them back to manageable height. When you do, you encourage the desired side growth of lateral branches. Always prune back to an outside bud to ensure bushier growth.

The third spring is the critical pruning time, generally speaking. What you do now is what you do each year to keep bushes productive, vigorous, and untangled. Cut off about one-quarter of the length of the canes that grew the previous season. This should leave canes 24

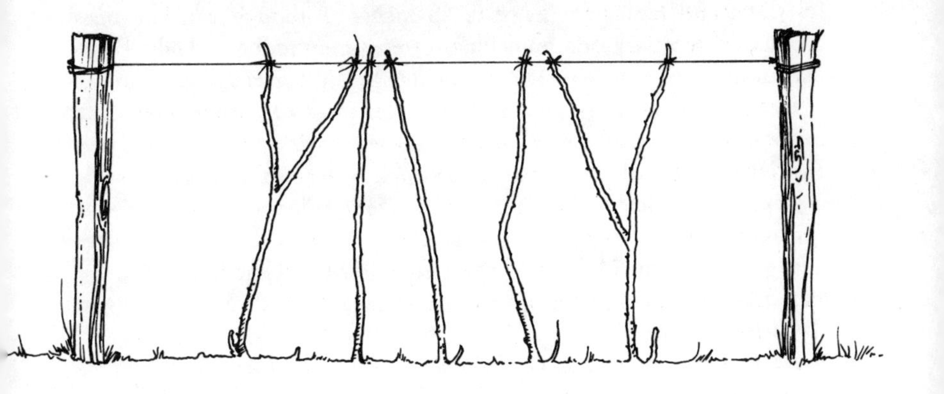

TRAINING AND PRUNING — HEDGE ROW SYSTEM

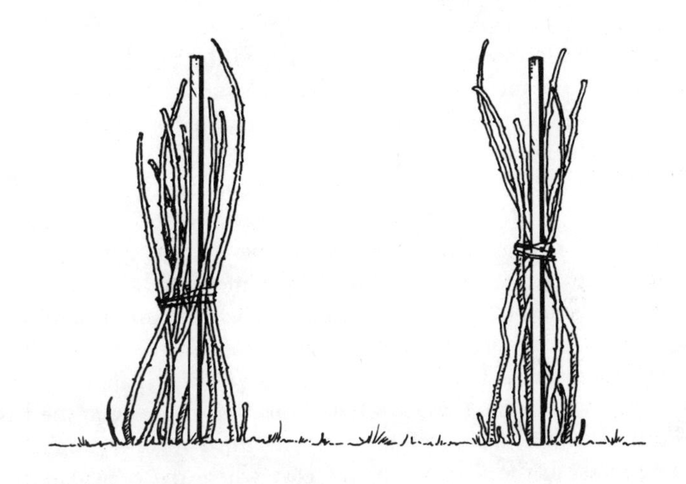

TRAINING AND PRUNING — HILL SYSTEM

Here are two ways to grow and prune red raspberries. The upper view shows them in the hedgerow form. The lower view depicts the hill system.

to 36 inches high. Some varieties grow taller naturally. Top height for them and for convenience is 48 inches. Although pruning does encourage necessary side branching, overpruning reduces yields. Raspberries must be trimmed, but not as much as other fruits need to be.

Remove weak, thin canes and dead canes at the same time. Any canes that show evidence of disease in any form should be removed and burned at a distance from the raspberry bushes. Pruning time is a good opportunity to look for any cankers, galls, or other problems. If you find any, check other canes nearby. Remove any afflicted ones and destroy them. Never use them for compost. That holds true for any diseased leaves of any plants, since some diseases linger even in the heat of active compost piles.

Raspberries are biennial. New canes appear, mature, and become next year's fruiting wood. After that they die and should be removed. Fruit is borne on second-year wood, so allow that and new young canes to continue growing. Thin these young canes only enough to keep plants reasonably open to sun and air and to prevent overcrowding. Remove all old, dead canes unless you need some for support of the plants in windy areas.

Black and purple raspberries should be treated differently in pruning, since their growth habit is different from red varieties. The blacks and purples should be summer topped during the first season to promote side branching. Topped plants also make harvesting easier and reduce wind whipping damage. Canes may sprout rapidly on good soil, so more than one summer pruning may be needed.

Remove the top 3 to 4 inches of growth when the cane has reached a height of 24 to 36 inches. If plants are less vigorous, cut canes at 18 to 24 inches. Don't prune in late summer or you will encourage new side shoots that are easily damaged by winter weather.

As you inspect your black and purple raspberries during the several spring pruning trips, also remove dead or diseased canes as you would with red varieties. Cut out weak canes too. Thicker canes are more productive, so eliminate those less than ½ inch in diameter at the base. By the way, when *diameters* of canes are noted for any raspberries, the term refers to the diameter at the base where the cane rises from the soil. That's standard terminology.

After topping back, prune laterals of the main canes 8 to 12 inches in length on black varieties, 18 to 24 inches on purple ones. The larger the cane, the more fruiting wood that can be left. Purple varieties produce more fruitful buds further from the main cane. That's why you should leave them slightly longer.

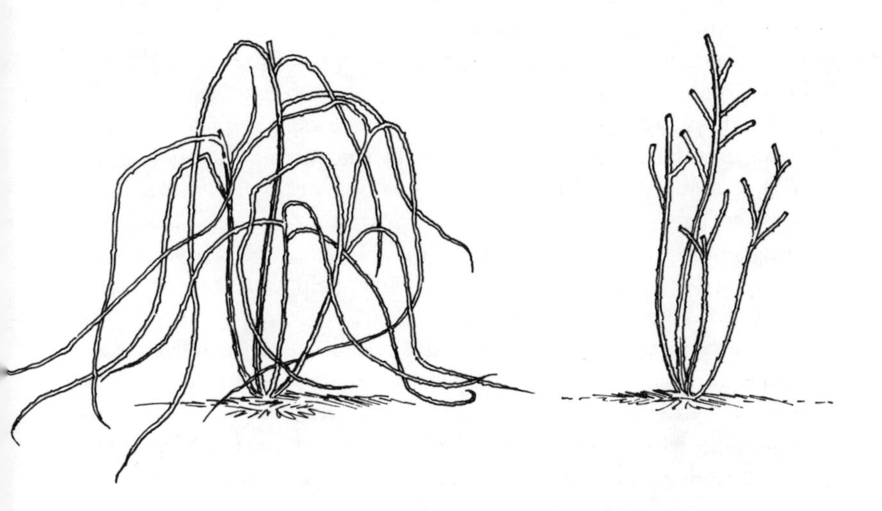

A black raspberry that looks like the one on the left may have a certain esthetic appeal, but it becomes more productive when it is pruned as is shown at the right.

The illustrations in this chapter show the basic methods of pruning both red raspberries and black or purple raspberries. It may seem to you that these illustrations show especially heavy pruning. Not really. All these illustrations have been checked by experts, including college fruit specialists and commercial growers who grow fruit for a living. They know how to get the most productivity from their land and best results from each bush or tree.

Raspberries will respond well to fertilizer and pruning. Sometimes they amaze you with the new crop of suckers they produce. If you prefer the hill system for black and purple, simply remove suckers that arise in other than the desired area. Digging is best, since breaking or pruning may leave the small roots intact to sprout again.

Between rows, we have found that a rototiller does a dandy job. We cultivate several inches deep and as wide as we prefer to walk easily. That eliminates suckers that tend to fill in between rows and make harvesting difficult. It also enables us to spread fertilizer and manure and do our pruning more easily than if the plants were allowed to form a matted thicket.

Raspberries need lots to drink when they bloom and begin to set their fruits. If you receive less than an inch of water each week, be sure to supplement it to produce those juicy berries. To measure the water you apply, set straight-sided cans or cups in the rows. When an inch of water has collected in them, that's enough that week.

If a drought hits hard, a second watering each week may be needed to replace the moisture lost down into the subsoil. Raspberries feed with shallow roots, so that top layer of soil needs the water. Soaker hoses are best. Avoid sprinkling late in the day. Excess moisture on foliage and fruit can cause mildews and molds. If these problems occur, a late fungicide application may help, but preventing the problem is wiser.

Be sure to water raspberries. This point deserves emphasis. Other fruit plants may still bear plump berries with moisture received from lower soil levels. Raspberries really can't do that well. Just a few days of hot, dry weather can shrivel the berries to the point where they are almost useless. They're such a treat, it's a shame not to keep watch over them at fruiting time to provide that little extra water that will let the plants produce to perfection.

PROPAGATING PAYS

Raspberries are easy to reproduce. Red varieties do it themselves. New suckers sprout all around the plants and along the rows. You most likely will need to remove some anyway in your thinning program. Just dig up the extras that you don't need for next year's fruiting wood and replant them elsewhere. Avoid using any that are weak or thin-caned or that show any signs of disease. Usually each parent plant will produce five to six new plants each year to let you increase your plantings. Transplant any suckers immediately after digging. The best time is spring, but early fall is satisfactory if you mulch around them the first year.

Black and purple types are propagated differently—by tip layering. The time is in fall when tips of the current year's lateral canes appear "rat-tailed" or with small curled leaves at the tip. Just tuck these tips into the soil at a distance from the parent plant. Cover each one well

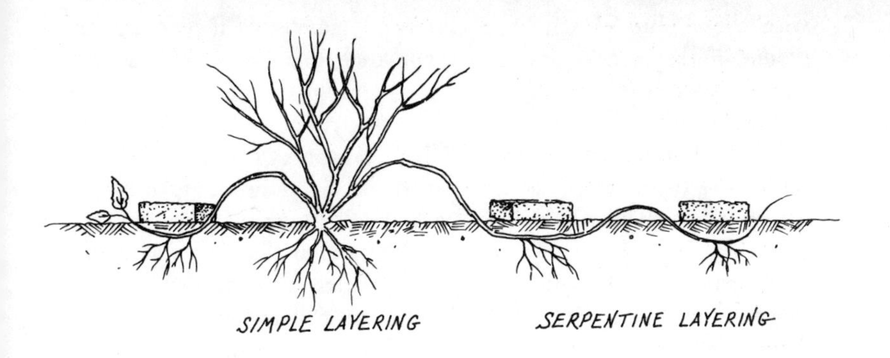

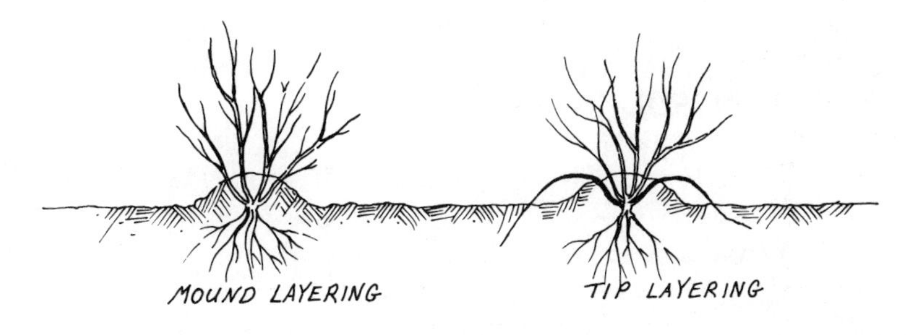

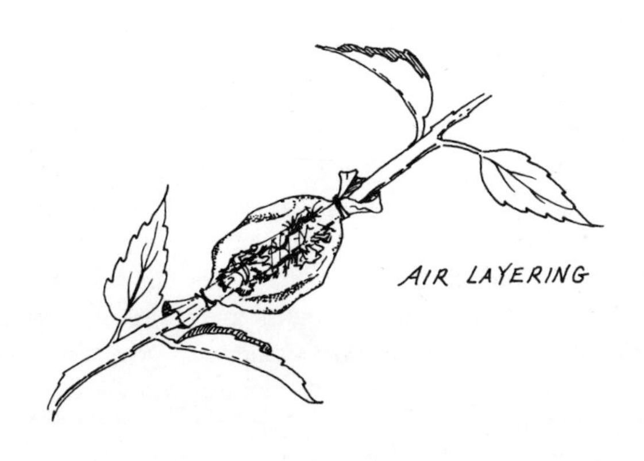

Different bushes in your yard can be propagated through several variations of a process known as layering; an easy way to many new plants.

with a shovelful of soil. You can bend laterals so they touch the ground if they aren't quite ready and tuck the tips in. Use a small stone or wire hook to hold a cane if it's springy and tends to snap out of the soil.

By late fall, the laterals are usually well rooted. That's nature's way with these types of raspberries. It is best to leave the new plants until spring. Cover with a mulch to mark them. In spring, sever the new plants from the parent, leaving an 8-to-10-inch portion as a handle. That makes moving and transplanting easier. Dig the newly rooted plants and set them wherever else you want to establish good growing raspberries. Then prune as you would with one-year-old stock brought from a nursery, and follow the recommended procedure year by year until new areas are bearing abundantly.

PEST CONTROL IS IMPORTANT

Insects can be a problem at times, but not as often as with tree fruits. If they occur, they are readily controlled by well-timed pesticide applications. It is important to control insects when they do invade because insects transmit plant diseases. On raspberries especially, plant diseases are more of a damaging threat than insects alone.

The special chapter on pest control outlines a recommended program to prevent insects and diseases on raspberries. Depending on the area in which you live, local recommendations may vary. It is best to consult pesticide suppliers or county agents for the latest programs suggested by agricultural college experts in your state that apply to home-garden fruit growing.

One interesting fact is worth noting. Insects have always preferred and still prefer to attack weak, less-than-healthy plants. This is true with both vegetables and fruits. The better you provide nutrients, water, and other care for your plants to keep them thriving, the better they can resist insects and diseases. That's true with people too. When you stay in shape, eat properly and rest enough, you have a natural ability to fight off problems.

HAPPY HARVEST TIPS

A well-established red raspberry planting after several years should yield 50 to 100 pints of tasty berries per 100 feet of row. That's mighty good eating. Black and purple raspberries yield somewhat less.

Raspberries are ready for picking, as Drew Swenson demonstrates, when they separate easily from the "core" without crumbling or mushing.

Picking time is important. Although red raspberries may last for a while on the bush, with some varieties it is best to pick when the time is ripe. Otherwise, fruit will deteriorate in just a few days and you can lose a large portion of your crop. That's even truer with black and purple types.

At their peak, berries may have to be harvested several times a week, especially if the weather remains hot. But if you provided favorable growing conditions, you'll pick plenty in that happy harvest season.

Here's how to tell when they're ready. Raspberries are perfect for picking when they separate easily from the "core" without crumbling or mushing. Grasp a sample berry firmly between thumb and two fingers. Pull with an even pressure. Hold only a few berries at a time to avoid crushing. If they pop off handily, taste juicily sweet, and have no oversoftness, you're right on time.

Move raspberries from picking basket to refrigerator quickly. Excess handling causes bruising and loss of quality. It is important to use pint or quart containers. If you pile berries in a deep container, the weight of the top berries will crush those below.

Harvest in the coolness of morning. Berries will keep for several days under refrigeration. If you plan to can or make jam, jelly, or preserves, do it as soon as possible. For freezing, you can put berries in pint containers and freeze them on the special freezing trays of your freezer. Another good way is to freeze them quickly on a metal cookie sheet.

Spread berries after washing to remove soil, grit, and dust. Dry them on absorbent paper towels. Then place them on a cookie sheet in the freezer. They'll freeze hard as marbles and then you can pack them more easily into larger containers for storage in your freezer.

Of course, do taste a few here and there in the process. That's the fun of having really ripe raspberries anywhere you like as part of your fruitful landscape.

11. Beautiful Blushing Blackberries

If you want to enjoy one of the tastiest home fruit berries and have a handsome hedge that thwarts wandering neighborhood pets, protects your property, and grows even in poor-quality soil, think blackberries. No longer are these merely the smallish, overtart berries found in bramble patches along a country lane. Today, blackberries are as plump as your thumb and juicier than ever. The bushes bear so prolifically that you'll have ample to use for jams and jellies, or even to sell in order to buy other plants for your home grounds.

Fact is, blackberries are so versatile they're enjoying a remarkable return to popularity. You can train them neatly as part of well-trimmed plantscapes or let them become an impenetrable hedge to screen unsightly views and protect property lines as well.

Blackberries are seldom sold in supermarkets or local stores. Occasionally you'll find them offered at country fruit stands. If you enjoy them, or just want a permanent hedge on your property, try the newest blackberry varieties. They're mighty prolific. On just two rows along one field, we harvested more than 300 quarts the third year after planting. The excess that we sold paid for lots of other trees and shrubs to increase our home fruit landscape.

If you prefer to avoid the bramble thorns that catch on clothes, you can do that too. New thornless blackberry varieties are available. They're not only thornless but also productive, vigorous, and winter-hardy to Illinois, Ohio, and New Jersey. Other types, with their typical thorns, thrive much further north.

Few fruits for the home ground are as dependable in production as blackberries. They prefer temperate climates, so they're not as well adapted to the Plains states or mountain areas. However, they thrive in such a variety of soils and perform so abundantly that they're worth a try in most other areas. Once established, blackberries will produce crops for fifteen to twenty years. That's surprising longevity.

Almost any type of soil is suitable for them, provided they can get from it (or you can provide) ample moisture come fruiting time. Adequate water then produces those succulent, plump, prime crops of berries. Blackberries also appreciate lots of sun, but can produce reasonably well in areas that are not suited for other berry bushes or fruit trees. Put their more tolerant growing abilities to work on some of these neglected areas.

You can select either the upright types or trailing varieties. The upright ones are superior. They are more productive, respond to cultivation well, and when well fed produce vigorous growth and bountiful harvests.

Typical growth is similar to that of their wild relatives. The erect bushes have arched, self-supporting canes. You can train them to arbors much as you would train grapevines. Several other systems work well. You can allow them to spread into a thick hedge to border a backyard. They can be guided to grow upright between two horizontal wires strung side by side between posts or beside two horizontal wires strung one above the other between posts.

Although blackberries can perform under a variety of soil conditions, it pays to treat them with kindness. They thrive best on sandy loams, since their roots may penetrate 2 to 3 feet deep. Clay soils may restrict this natural growth and consequently their future yields. Where shelter from harsh winter winds is possible, give them that advantage. Some varieties can take more cold than others, but if you have a windbreak of a house, garage, or tree, blackberries appreciate this consideration. Planting along a slope that gets good sun is helpful too. As with all fruits, avoid frost pockets or areas where water stands in winter.

With the renewed interest in blackberries, nurseries are offering several excellent, taste-tempting varieties. Their berries, thanks to the skill of plant breeders, are much bigger than any wild variety. Here are some of the best among the erect. Since they are so much easier to tend, and perform so much more satisfactorily, I've concentrated only on erect types in this chapter. Trailing types, more commonly called dewberries, are included with currants and other minor fruits later.

Lawton has been widely grown for its medium, sweet fruits. It is restricted to southern areas.

Raven has medium-large fruits, medium-firm with excellent eating quality. Plants are moderately vigorous and productive.

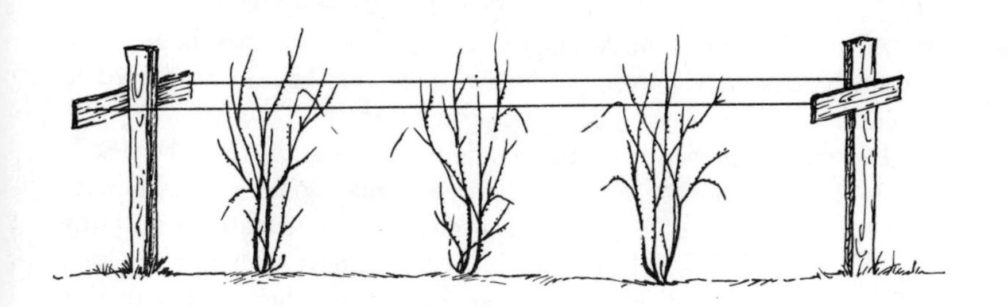

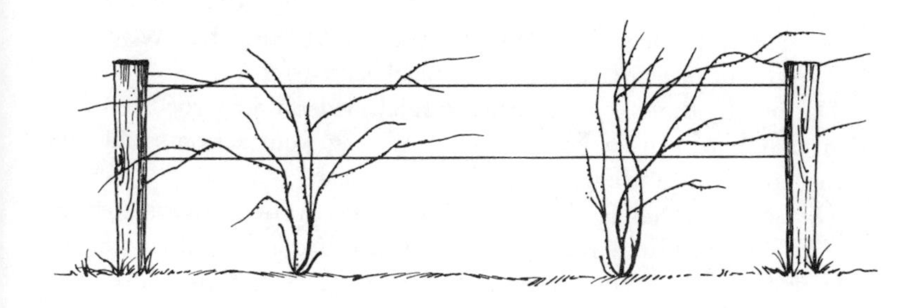

Here are two good ways to train blackberries. In the upper view, the horizontal wires are side by side. In the lower view, the wires are one above the other.

Ranger is similar to Raven but more adapted to Atlantic coastal areas.

Smoothstem is one of the first thornless varieties. Fruit is medium to large, firm, tart, and semi-upright. It can be grown on a trellis, but may suffer in hot, dry, windy areas.

Cherokee is a recent (1974) introduction from the University of Arkansas. Fruit is medium to large and firm with vigorous, erect canes.

Comanche is another large, firm new variety, similar to Cherokee but ripening early for home-garden use.

Blowers is a good midseason berry of dessert quality. It has large, firm berries on hardy vigorous bushes and ripens over a long season.

Early Harvest has medium-size, firm berries on moderately vigorous bushes. It is better suited to southern areas and is very productive in the south, as far north as Maryland and southern Illinois.

Jerseyblack is an early, long-fruit-season variety with medium to large berries. Bushes are semitrailing and need some support.

Alfred, developed in Michigan, is early too and has large, firm, sweet berries on vigorous, productive, hardy bushes. It is adapted to more northern areas similar to Michigan's climate.

Darrow has proved to be the standout of all the top-rated berries. It is so far superior to all other varieties that some nurseries have dropped other types in favor of Darrow. It is particularly noteworthy for its vigor, reliably heavy production, firmness, and quality, year after fruitful year. Plants are hardier than many other types, a quality that fits it well for northern states. Berries may range an inch or more long and ¾ inch wide and are glossy black. Another nice thing about this productive blackberry is its long harvest season. You may need to pick it several times, but Darrow continues yielding for weeks. It's fine fresh, for pies or freezing, or for blackberry wine.

Darrow, by the way, forms such a thick hedge under good fertilization that it served to keep some wandering ponies from galloping across our gardens when they got loose one year.

Dirksen Thornless and *Black Satin* are two varieties originated by the U.S. Agricultural Research Service in cooperation with Southern Illinois University. Both are thornless, productive, and vigorous. They aren't so hardy in northern areas by may be worth growing where warmer climates encourage it and a thornless bush is desired.

For details on trailing types, consult the chapter on currants (Chapter 12), which includes dewberries. They can produce well but in many cases require extra care in training, and their rambling habit does require lots of room if they are left on the ground. For most satisfactory blackberry growing, you're well advised to choose and use the erect types. They're generally hardier, more vigorous, more productive.

PLANTING POINTERS

Nurseries may sell rooted cuttings of erect-type blackberries. These would be 4 to 6 inches long, about ⅛ to ¼ inch in diameter. They don't look like much when they arrive, but they'll surprise you. The more vigorous rooted suckers, 12 to 15 inches long, are a better bet.

Prepare soil as thoroughly as you would for a garden. Although blackberries will perform on poorer soils, they welcome additions of organic matter to soil. Manure, if available, can be tilled into the soil the fall before you plant. That preparation helps set the stage for their prosperity. Since they do establish deep roots, digging or tilling 8 to 12 inches deep is worthwhile. If soil tends to be heavy, you can prepare slight ridges to receive the plants. That way, excess moisture drains away but is available for feeder roots as the bushes mature into hedges in future years.

You can plant blackberries anytime during the dormant season, but early spring is best. That way, you won't shock tender varieties if the winter is especially severe that first year.

Blackberries need somewhat more room to ramble into natural brambles than do other types of berries. You may not think so, looking at the short rootstocks that arrive from the nursery. However, blackberries are noted for another amazing habit. They propagate themselves quite rapidly with underground runners or suckers.

If you can't plant immediately when the rootstocks arrive, heel them into the soil as you would other trees or bushes. Keep roots moist. Planting on a hot spring day, one year, I lost a fair number of the plants because they dried out in the sun. To avoid that hazard while working in the garden, keep moist burlap over the roots. Better yet, keep the plants in a bucket of water with moist burlap over the canes. They may be somewhat touchy as you plant them, but once established, they'll be with you for years to come.

Not so incidentally, if you have wild blackberry bushes on your property, try to eliminate them before you introduce cultivated varieties. These native wild ones may be resistant to disease problems and harbor some infections that can be transmitted to domesticated varieties. That's a good approach with other fruit crops too. Often wild types can survive with no apparent ill effects but can transmit problems to the types you plant. So it's best to eliminate the wild ones first.

If planting the small root pieces 4 to 6 inches or so long, set them about 6 inches apart along the row. Most nurseries sell the more desirable rooted stock that is dug from one-year-old suckers in the nursery. These are a better buy. They are stronger, take root faster and begin to send out their own underground suckers the second year to fill in the rows. These rootstocks may be 12 to 18 inches tall. Set them 3 to 4 feet apart in rows. If you prefer parallel rows, keep the rows 6 to 8 feet apart. Blackberry plants spread along the row and fill in between them, but you'll want room between rows to tend them, so follow this spacing guide.

Set plants the same depth at which they grew in the nursery. You will see that point on the canes, just above where the roots have formed. Tamp the soil around rootstocks well to establish firm contact with soil. Then water. A good soaking is helpful at first and again each week until they get well started.

If you desire just a clump or grouping in a corner, you can place plants 3 to 4 feet apart. They'll fill in well in later years. Judicious pruning will let you walk among them for harvesting. For most effective use of superior varieties like Darrow, hedgerow planting is recommended. It costs a bit more to space plants closer, but you'll find they fill in more rapidly, especially if mulched with straw and manure and fertilized each season.

After planting and watering blackberry plants, take your pruning shears and cut back to half the height from the ground. This encourages the desired side growth of lateral branches.

FEEDING BLACKBERRIES

Mulching blackberries with compost, straw, grass clippings, and similar materials is better than clean cultivation. If you don't have enough mulch for these bushes, cultivate them just a bit to remove grass and weeds. Blackberries are shallow-rooted; you shouldn't disturb them as they set their first year's vital root structure.

Although blackberries do thrive on poorer soils, it also is true that they respond quite vigorously to soil improvement with organic material. If you can provide a winter cover mulch the first year, do so. Some varieties, you'll recall, are particularly prone to winter injury. That first season is the toughest on them until they are more deeply rooted.

For maximum yields, apply fertilizer every year at blossoming time. The year we used several inches of stable manure was the best one ever. Unfortunately, that may not be available, or other manures either. If not, apply a commercial 5-10-5 as a top dressing along the rows each year when they bloom. Spread an average of 1 cup per parent plant along the rows, ½ cup per side, and water after you apply it. That parent plant has probably grown youngsters, and extends 6 to 8 feet along the row. For long rows, figure 5 to 10 pounds of

5-10-5 per 100 feet of row, depending on maturity and density of the rows.

Judging from split fertilizing trails we've seen, I believe it pays to apply half the fertilizer just before blossom time to get canes and new fruiting wood in prime condition. Then apply the second half within several weeks as tiny fruits are forming. As soon as fruits form, pay more attention to the water. Blackberries especially respond to copious amounts of moisture to achieve their plumpest, sweetest size for you.

PRUNING IS PARAMOUNT

If you wish to control blackberries, pruning is necessary. If you wish them as a bramble patch with appearance closer to their more natural state, just periodic trimming to let you walk among the plants is satisfactory. Blackberries astound most home gardeners with their constant spreading habit.

Crowns of blackberry plants are perennial. New canes arise from them each year. The canes, however, are biennial. They live only two years. During the first year after planting, let your plants bush out. The second year, you'll find new suckers filling in the open spots and new canes sprouting from the crowns and around them too.

During the first two years, blackberry canes send out laterals, also called side branches. The second year, small branches grow from buds on these laterals. This is the fruiting wood. After the laterals fruit, these canes die. That makes pruning an easy-to-see, simple procedure.

Prune laterals back each spring after the second year. Before growth starts, cut laterals to about 12 inches long. Check the ground area for new sucker canes arising. If you let all grow, they would turn your bushes into a matted thicket.

During each growing season, remove all suckers that arise between rows or too far from where you wish your plants to stay. Use heavy gloves and pull up these out-of-place suckers. If you merely cut them, they often regrow.

When tips of erect blackberries reach 30 to 40 inches, cut off their top tips. This pruning forces canes to branch. Tipped canes also grow stouter. They're more capable of supporting heavier fruit crops.

Come summer, when your harvest is over, cut out the old canes.

They're easy to distinguish—in both size and color—from the new wood, which will produce next year's fruit. If too many new young canes have grown, thin them back, leaving four to six per bush. It pays to burn old canes. Anthracnose and rosette can be problems, particularly in southern regions. Burning eliminates this threat, since the disease overwinters mainly on old wood.

The two-wire trellis will neatly hold your taller canes from drooping. In spring you can easily place those out of position within the desired guidelines. This support also helps keep your fruitful crop at a handier picking level.

After you've removed old canes, trim lateral branches back as you did the year before. Cut them to about 12 inches in length. The number of canes per foot of row may vary according to variety, but practice will guide you year to year.

Blackberries can become exceedingly tall if not trimmed back in fall or winter. Whether you wish a bramble hedge or not, remember that pruning does encourage greater yields. You can sacrifice some yield by letting them grow thick, but increased fertilizer, especially manure, will create a veritable forest of canes. Few animals enjoy pushing through the thornier varieties. That's what makes them ideal for property lines. But when they are overdense, yield suffers.

If you prefer a trellis pattern of more decorative effect, you can tie canes in bundles. Then attach them with stout cord to the desired wire, wood, or fence supports.

HARVEST WHEN DEEP BLUE-BLACK

Blackberries received their name on purpose. Too often we may be tempted to pick them when they seem plump and dark, but just a few tastes will correct that error. As fruits mature, they become more intensely blue-black. Berries increase in size and become less firm. As they seem to be ripening, pluck a few. You'll soon realize when they're fully ripe. Some varieties have extended harvests. Darrow, the best of all the most productive ones, does provide a longer picking time. Look them over every day or so, and pick the ripe ones.

Pick early in the day. Blackberries won't spoil as rapidly if picked when it is cool as they will if you wait until the sun if hot and bright.

PROPAGATING IS FUN TOO

Blackberries not only are abundantly fruitful but also provide dozens upon dozens of new sucker shoots each year. You can profit from this habit too. Once your neighbors have tasted these tempting, juicy berries, they'll probably want to swap you for some of their different garden plants. Why not trade? Just dig up the sucker plants, those about 12 to 18 inches tall, roots and all. Keep them moist and cool until they can be replanted.

You can encourage even faster reproduction in your own garden rows for trading or even sale to others. Bend tips of the tallest canes to the ground. Cover each with a shovelful of soil. By next spring, those canes will have rooted. Just separate them from the parent, dig up roots, and you'll have true-to-variety new plants to use as you wish.

12. Keep Current with Currants, Dewberries, Gooseberries, and Such

In jolly old England, which I visit periodically to keep up to date on European gardening ideas, currants and other tasty old-fashioned fruits are commonly grown. These delicious and easy-to-grow fruits have never really taken root in America. There's no good reason they shouldn't. Many of these less-known fruits can be a colorful addition to home plantscapes. They take little room, and some can be trained on posts or fences. Others sparkle with blooms as interplanted bushes with other shrubs and trees.

By summer, currants also provide a spectacular display of brightly colored fruit for jellies and jams that are a different dimension in tastier living from your home garden.

One problem that has restricted home growing of currants and gooseberries has been certain insect and disease problems associated with them. Federal and state regulations have prohibited movement and in some cases the actual growing of these plants. These restrictions were thought necessary in order to prevent transmittal of any exotic plant pests that might become a threat to other major commercial crops.

Fortunately, because of new varieties, clean rootstocks, and improved materials to fight insect and fungus disease, currants, gooseberries, and other unusual fruits can be grown widely. Restrictions still do exist in some areas.

You may wish to check with your county agricultural agent or state horticultural extension specialists at your state college of agriculture about this situation for your area. A list of the key colleges in each state is included in Chapter 20. Extension specialists and county agents are good people to know, for many reasons. They are constantly updated on new varieties, growing techniques, and improvements in pest-control programs that apply to your local area.

Be aware of another important point. All plant nurseries, whether just propagating and selling within one state or shipping their trees and shrubs throughout the United States, must comply with certain basic rules. Plant health inspectors from the various states and the U.S. Department of Agriculture inspect growing stock in nurseries. They check for any insects or diseases. If they find them, usually that particular stock must be destroyed or corrective measures must be taken at the nursery to eliminate the pest problem and bring plants back to perfect health.

These are good rules. They protect all of us as consumers so that plants we buy don't introduce problems into our home grounds. After all, we face enough already without having new troubles arrive with a shipment of trees or shrubs.

Currants and gooseberries are delights. Once you have enjoyed some black and red currant jam or jelly on crisp, hot toast, you'll have some idea how good these fruits can be. Home-grown, they're even better. Currants and gooseberries are quite hardy and easy to grow in home gardens.

Although you can buy plants of these two fine fruits that are perfectly healthy, they may serve as alternate host to the white pine blister rust disease. Some states require a permit to grow currants and gooseberries and won't issue one if you live within 1,500 feet or so of sizable planting of ornamental or commerical white pine stands. To be a good gardening neighbor, do check on the restrictions in your area. You don't want to encourage problems in nearby woodlands. You can expect land owners and forest companies to be quite touchy if your currants become a source of this problem in their trees.

Currants and gooseberries are used mainly in making jellies, jams, preserves, and pies. Red gooseberries are sweet when fully ripe and may be eaten fresh. Currants may be eaten fresh, but they may be a mite too tart for you.

Wilder is one of the best currant varieties. It has large, dark red, sub-acid berries that hang in large compact clusters. The bush is upright and large. It may grow 4 feet tall or more without pruning, but is easy to pick and is vigorous.

Red Lake ripens after Wilder. It has large, firm, lighter red berries. Clusters are large and hang on long after ripening. It is upright, vigorous, very hardy, and nicely productive.

White Imperial is a midseason variety on a small, spreading bush.

Clusters are medium to long, well filled with medium to large, creamy-white, sweet, juicy berries.

For northern areas especially, *Minnesota 71* is a spreading vigorous variety that bears well-filled large clusters of plump, good-quality berries in midseason.

White grape currants, the variety usually sold by nurseries, is similar to White Imperial but is superior.

Perfection ripens in midseason. It is fairly productive with large, bright crimson berries on a somewhat spreading bush.

Some nurseries offer black currants, but they may be difficult to find. Order a variety of nursery catalogs. Look them over. Then, pick those varieties that seem to fit best in your planting for size, shape, bearing ability, and hardiness.

GOOSEBERRY VARIETIES

Although European varieties of gooseberries are larger, American varieties are more productive and hardier and are considered tastier and of higher quality. Just as important, they do better in our climate and growing conditions.

Downing bears large, pale green fruit. It is probably the most widely grown variety and is preferred for canning quality, which is excellent.

Glendale has medium-size, dull-red fruit. This plant grows rather large and is vigorous and productive. Since it withstands heat best, it is recommended for southern areas.

Fredonia is moderately vigorous and productive. Berries are exceptionally large, dark red, and attractive. This is probably the best English-type gooseberry as well as being one of the largest available.

Poorman belies its name. Plants are the most vigorous, healthiest, and most readily productive of varieties that perform well in the northern parts of our country. Bushes are the largest of any American type. Berries are red, attractive, and of the best quality. This variety is tops and deserves a place in your home garden.

Chautauqua is medium-early and has very large fruit for dessert use. The fruit is greenish yellow, the bush small and somewhat spreading. It may fit better into some of those special spots where taller, more vigorous bushes would be crowded.

PLANTING-TIME TIPS

Currants and gooseberries need a cool, moist, and somewhat shady location. That's handy. You can grow them in spots not suited for other fruit crops. Since they are bushes, they'll fit into property borders or as groupings in corners, or do well interplanted with other flowering shrubs along a shadier side of your home. You can use the taller growing types along a wall of your garage or home. Then select the lower-growing spreading varieties in front of the tall ones. That way you have a wider selection of plants, plus a range of taste treats from their natural differences.

Currants and gooseberries are quite resistant to low temperatures. But with the exception of some heat-tolerant varieties, they don't thrive where summers are hot and dry.

Gooseberries in general are somewhat more tolerant to heat than currants are. If you want to try both types of fruits in southern areas, select the north side of a building, which receives abundant shade. That can help protect these plants from summer's heat.

Pick a site with good air circulation and well-drained soil. You can, of course, improve any soil before planting time. Details of these simple, effective steps are in Chapter 2, Good Earth Basics.

Currants and gooseberries both bloom early in the spring. That means they may be damaged by late frosts if they aren't protected. You can erect wind screens of stockade fencing or of burlap fastened on poles. Both currants and gooseberries are shallow-rooted. Although they require moist soil, especially as they set their fruits, to help them achieve the plumpest, juiciest size, neither currants nor gooseberries can stand wet feet. They grow best in deep, fertile loam in a pH range of 6.0 to 8.0. Although you can be successful with these plants on lighter soils, they really prefer heavier types such as silt or clay loams. Adding abundant amounts of organic matter to a sandier soil increases its water-holding capacity. Drainage is important, but adequate supplies of moisture are even more important to currants and gooseberries.

Vigorous one-year-old plants are the best buy. When shopping for fruit trees or plants, remember this: younger ones are not only less expensive but also easier for you to train and prune to the desired shapes. Since pruning is so vital a part of successful fruit producing, it is usually best to start with vigorous young plants and train them as you wish.

For a berry patch, you should plant currants and gooseberries about 4 feet apart in rows 6 to 8 feet apart. However, if you prefer specimens interplanted with other shrubs and bushes, plan for growing room about 4 to 6 feet in diameter, depending on the variety.

In Chapter 2 you'll find an excellent illustration of the basic planting methods for fruit bushes. It applies to all shrubs and trees, even more so to those that require good drainage.

In this chapter, we're concerned with the specific requirements of currants and gooseberries. Set them slightly deeper in the ground than they grew in the nursery. This is important. It will cause them to grow new shoots from below the soil level. The new shoots will form bushes, rather than single or merely double-stemmed plants. Pack soil firmly about roots and then water well.

After planting, cut tops of currants and gooseberries back to 8 to 10 inches. This pruning encourages side branching initially to provide more desirable bushier growth patterns. The bushes will look better as a landscape plant, and this practice induces the plants to be more fruitful in line with your multipurpose plan.

Mulching is a sound practice on these plants. Any good organic material is fine: grass clippings, compost, peat moss, or leaf mold. Or you can try black plastic. Spread organic mulch in a 3-foot circle around each bush. Before winter, however, be certain to pull the mulch back from around the bushes. That move will eliminate nesting spots for mice, which are fond of feeding on young shoots.

Currants and gooseberries respond to fertilization even when growing in already fertile soils. Plan to make an annual fall or late-winter application of well-rotted barnyard or poultry manure. Spread this about 1 inch deep in a 3-foot circle around each plant, right on top of the mulch. As it rots down, it will help decompose the mulch, adding vital nutrients to the soil. These plants enjoy high organic matter in the soil, the more the better. No soil can get too much organic matter.

You can, of course, use commercial fertilizer. If you do, apply 1 cup of 10-10-10 around each plant after planting and again each spring before the buds break forth in bloom. On sandier soils you should continue to build up organic matter with manure, straw, or compost each year. It also pays, on sandier soils, to increase the fertilizer rate somewhat, since a bit will be leached into lower levels and lost to the feeding roots. Apply 1½ to 2 cups per plant of 10-10-10 or even 10-6-4 left over after lawn fertilizing in the spring.

If you don't have enough mulch material, your next best bet is clean cultivation. Weeds compete with all plants for moisture and nutrients. Currants and similar bush fruits, in order to produce abundantly, need nutrients, so keep weeds away from them. If you cultivate rather than mulch, just scratch the soil slightly. These plants have shallow roots, and you don't want to disturb them by deep cultivation.

Gooseberries and currants form bushes with many branches rising near the original plant. Do your pruning anytime during the dormant period (when leaves are off). Pruning these bushes means thinning out excess stems. It is really quite easy. Snip away any rubbing branches and broken or weak stems. Shape the bushes to fit the look you want in your landscape scene. Little pruning is needed until plants are four years old.

At maturity, the typical mature bush should have three or four stems each of one-, two-, and three-year-old wood. Count that again. It means three or four stems of wood of each of those years' growth. The actual number is determined by the vigor of the bush. Since these bushes are low-growing, only occasional top pruning of taller varieties is necessary to keep plants within their allotted spot.

Remove all wood over three years old. You can identify old wood by its older, darker, more weathered appearance. Remove any prostrate stems. Wood over three years old is less productive than young, vigorous wood. Since the best crops are borne on two- and three-year-old fruiting wood, you'll want to encourage it to continue forming from the most vigorous one-year-old branches.

WHEN PICKING TIME ARRIVES

Currants and gooseberries will begin bearing the third or fourth year. They have a productive life of ten to twenty years under good culture. You can expect 5 to 15 quarts of berries per bush from currants. Gooseberries are normally even more productive, yielding 10 to 20 quarts or so per bush.

Other fruits should be picked as they ripen. Currants and gooseberries may be left right on the bush for several weeks after they ripen. When you do pick, avoid bruising the fruit. Gooseberries may sunburn or sunscald if you leave them in the sun after picking. (I know you won't do that, but I thought you should know about that little problem anyway.) Both fruits can be kept in the refrigerator for a time; but since they do ripen fully on the plant, putting them immediately after picking to their intended use in jelly, jam, or preserves is best.

ELDERBERRIES

Nothing else tastes so fine as elderberry wine. Way back when I was wandering the wetter portions of our farm fields in the Garden State of New Jersey, my job was to pick the wild elderberries each summer. All I ever tasted was the jam. My older relatives favored the wine, which I in time found one of the more exotic beverages among all homemade fruit drinks, applejack notwithstanding.

The wild elderberry of field and fence rows has always been popular for pies and wine. Today, thanks to those ever-alert plant breeders, elderberries have been well domesticated. Unfortunately, few people know or appreciate them; yet they can be a truly unusual part of your fruitful landscape.

Elderberries are nearly self-unfruitful, so you'll need two different varieties planted near each other so that they can cross-pollinate to ensure a good fruit set.

They like sunny locations where there is, or they can be given, ample supplies of moisture. If you have a little stream or boggy area, that's the best spot for them. They should have good drainage, but lots of moisture is vital to their livelihood. These bushes grow rather tall, about 4 to 6 feet, and have a spreading habit. You'll need a spot with 8-to-10-foot diameter to accommodate the two bushes needed to pollinate each other.

Adams is an old variety. *New York 21* has Adams as a parent. New York 21 has berries larger than most named varieties. It is somewhat smaller in its bushy growth but highly productive and bears earlier than many others.

Nova was developed from *Adams* #2 and was introduced in 1960 by the Kentville Experiment Station in Nova Scotia. Bushes are large. Nova is one of the most productive varieties available. In trials comparing it with others in Ohio and elsewhere, Nova exceeds all other elderberry varieties in yields. It ripens quite early in the season.

York is more productive than Adams varieties and is larger and bushier. Early tests have been substantiated that this newer variety is more productive and larger-berried than many others. It ripens later than Adams types.

When you select varieties, of these and other fruits, remember that different varieties may bloom at slightly different times. If you wish to stretch the harvest season, select early, midseason, and late varieties so that blooming periods overlap for effective cross-pollination.

Look for a site with reasonably well drained soil, but one that can provide, by natural rain or nearby stream or artificial means, enough moisture to let berries form plump and full. A sunny spot encourages strong growth year after year.

Pruning is simple. Just keep bushes trimmed within their allotted space. They'll spread nicely and droop their top branches when filled with berry clusters of bright purplish or bluish purple. Since birds like elderberries, it pays to use special netting over them as they approach maturity.

DEWBERRIES

Dewberries, called boysenberries in some areas, have a trailing habit and will crawl along the ground, down a slope, or even up a pole or trellis with a bit of help. They've never become a popular home-garden fruit, but a bit of information may encourage you to sample these appealing fruits.

The dewberry is a relative of the trailing blackberry. The youngberry, boysenberry, and loganberry are cultivated hybrids of the trailing dewberry. They trace other parts of their heritage back to crosses with blackberries and raspberries.

Dewberries can grow on several different types of soil, but they prefer well-drained soil with a clay or heavier loam subsoil. You should prepare soil thoroughly and incorporate organic matter from old leaves or compost into it as you dig or till. That helps hold moisture, but not too much. Pick a sunny spot if possible, although these ramblers can be used in slightly shadier areas if they get six hours or so of sun a day.

If you have room for them to roam, set plants 6 feet apart in late winter or early spring. Plant them as you would bramble plants like the better-known trailing blackberries.

Be sure to set the bud straight up and leave old vines on so you can locate the plants during cultivation before vines begin growing again. But don't cover the bud more than 1 inch deep.

Boysen and *Young* varieties are good for home gardens. Both are trailing types. *Flint, Williams,* and *Thornfree* are classed as semi-trailing.

Lucritia is an early berry of good quality. It is large, long, and firm but is susceptible to anthracnose and leaf spot. *Boysenberry* and *Youngberry* are midseason, vigorous, and moderately productive. They bear large, reddish-black berries of good quality. Canes may be killed right to the ground in northern areas if not protected by good straw or hay mulch cover during the winter. You might also mound soil over plants after removing vines once they have finished bearing. In some specialty fruit nursery catalogs, you can find others listed that have proved satisfactory.

As soon as vines have begun growth in spring and are 1 to 2 feet long, add ¼ cup of 10-10-10 around each plant. About a month or six weeks later, add another ½ cup of fertilizer around the plants in a circle 3 feet across. If you plant rows, you can side-dress the rows with similar amounts. These plants enjoy eating well. Consequently, another post-harvest feeding with ¼ cup per mature plant can keep them thriving.

Dewberries will need training, whether on the ground or up a fence. Simply tie the vines up the post and along the wood or wire trellises. You can also tie individual runners together in bundles and attach them with stout cord to upright supports.

Be careful with cultivation. These plants are shallow-rooted. Just remove weeds unless you prefer to mulch, which is really more desirable. Mulch retains soil moisture and provides a cushion beneath dewberries if you have let them ramble on the ground.

These fruit plants require little or no pruning on either cultivation system. But when harvest is over in the summer, cut off all the vines, new or old, and dispose of them. That step eliminates winter homes for troublesome pests, including vine borers, that may bother these plants occasionally. New vines will grow next spring, more quickly than you realize.

If you discover that dewberries please your palate, it's easy to produce more. Simply cover tips of growing plants with a handful of soil in late summer. Roots will sprout on these cane tips by spring. Then, snip off the old vines a foot away from where it rooted and dig up the newly rooted plant to start more dewberries growing elsewhere.

13. Try True Blueberries

Big, bright, and true blue, that's blueberries. Try them in pies, on cereals, in muffins, or fresh from the bush. You can do so when you plant those remarkable blushing blueberries as part of your delicious plantscape.

Blueberry bushes in the wild have two distinctive shapes. There are the high-bush and low-bush types. Forget about the low-bush kind. They are difficult if not impossible to cultivate and just don't yield anywhere near the abundance you can get from the prolific high-bush blueberries. If you must gather low-bush berries, a walk in the woods during vacation along swampy areas may yield a basket or two. For practical purposes around your home, the high-bush berries are your best bet.

Fact is, blueberries are one fruit crop that can give you the greatest pleasure and the most fruit to eat for the least effort in the backyard fruit garden. They provide pleasure in three ways: (1) Blueberries serve well as ornamental shrubs; (2) they're a productive source of food for you and your family; and (3) they're also attractive to songbirds. That last fact may prove more a minus than a plus unless you protect your crop before the birds come to sing for their supper.

Blueberries are somewhat different in their soil needs from other fruit bushes. Blueberries prefer soil that is loose, well aerated, and acid. The pH should be about 4.8 to 5.0 for best results. All other fruits prefer a pH of 6.0 or slightly higher, which is a good bit sweeter than the acid conditions blueberries enjoy and really must have to thrive. That's no real problem.

Most homeowners who live east of the Mississippi (where blueberries are native, grow best, and are long renowned) realize that soil tends to be on the acid side. That's the reason lime is recommended to improve growing conditions for lawn grasses that prefer sweeter soil.

See, you most likely already have soil leaning on the proper side of the pH scale for blueberries.

The factor known as pH, by the way, is a gauge that tells whether soil is acid, neutral, or alkaline. The lower numbers indicate acid condition. Numbers from 6.0 to 7.5 indicate neutral or so-called sweeter soil, but higher numbers (8 to 9) indicate tendency toward alkaline conditions. Simple soil tests by your county agent or local garden center or with a home soil-test kit can tell you quickly where your soil ranks on the pH scale.

For blueberries, 5.0 is good. You can make soil more acid in several ways and quite simply. The addition of oak and maple leaves, dug or tilled under, as well as pine or evergreen needles, tends to acidify soil. Adding nitrogen fertilizer also helps. Wood chips and pine bark especially build up the acid level. Lime, on the other hand, neutralizes acidity and sweetens the soil. For blueberries, unless soil is extremely acid already, you won't need even to consider the use of lime. In fact, keep it away from your blueberry planting area.

Blueberries are different from most berry bushes in another unusual way. The blueberry root has no root hairs as do many other plants. However, the entire roots system is very fine, fibrous, and hairlike in structure. That's a plus and a minus.

Such fine hairs just can't push their way through heavy clay-type soils. A well-aerated soil on the sandier side is needed. That's important to know as you select your planting site or begin improving the soil that will support your blueberry bushes. It must be open and porous to provide the minute passageways for easy movement of plant roots. Sandy to loam soils are ideal. It is important to have or incorporate high amounts of organic matter into the sandier soils. Blueberries need lots of water.

When you see blueberries in the wild, you'll notice they thrive best on highly organic sandy soils near streams and around ponds. However, they can't abide wet feet, since constant water in the soil clogs air space and will rot their roots. That's true with most plants, but especially true with blueberries. Because they are very shallow-rooted and have such fine root structure, adequate moisture must be available at all times and especially at fruiting to produce the plumpest, tastiest berries.

If your chosen site needs more organic matter, plan at least a year ahead before planting if possible. Add manure, straw, peat moss, compost—as much organic material as possible. You can provide more from year to year in the future with mulches that do rot down bit by bit. However, blueberries get off to a strong start when the area is well prepared. Avoid sod areas if possible. You are well advised to dig in organic matter and keep the area tilled for one season before planting blueberries. If more organic material is available, spread it on the area as mulch. Add pine bark or peat, which you can get from garden centers.

If pine needles and oak leaves are available, add them too. Let the material decompose naturally, just as you would in the natural layering method of making compost. In fact, consider your blueberry area a compost bed. The fall before you plan to plant blueberries, till or dig the organic material under. It will rot even more as it becomes part of your improved soil mixture for your new plants.

Natural blueberry soils in commercial blueberry-growing areas of New Jersey, Wisconsin, Rhode Island, Michigan, and parts of the South from Alabama to Texas have a naturally high water table. Water is seldom more than 12 to 18 inches below the surface. That's a key to good blueberry culture. If you don't have a high water table, keep in mind that consistent mulching and watering will be necessary for best blueberry production. That's easy to do with a soaker hose attached to a timer. You turn it on for fifteen minutes each morning and it shuts off automatically after giving your blueberries their necessary drink. The mulch prevents the water from evaporating.

Blueberry varieties are available to fit your climate. In northern areas the *Vaccinium corymbosum* and *Vaccinium australe* and hybrids developed from these types are quite hardy. They thrive from the Carolinas to Michigan and Maine.

In southern areas, another group descended from *Vaccinium ashei* has been improved upon by plant breeders. It is a high-bush type better known as rabbit-eye blueberry across the South from Florida and Alabama to Texas. These plants can soar 10 feet tall, but they can be kept in more manageable bounds by pruning. The northern types are usually 3 to 6 feet tall. These, too, need pruning to maintain their shape and promote more abundant bearing.

All types of blueberries have the typical hairlike fine root filament system. That's another reason you should prepare soil for them a year in advance. It gives you an opportunity to eliminate persistent weeds such as Bermuda grass in the South and nutgrass and other pesky weeds in more northern areas. Weeds rob soils of an amazing amount of water as well as nutrients. Blueberries rely on ample moisture near the surface to thrive. Weeds must be banished, by removing them or smothering them with mulch.

Mulch is, of course, my preference, especially on such shallow-rooted plants as blueberries. Disturbing their fine roots can damage and stunt or even kill your plants. Besides, the more peat moss and sawdust you add, the better you maintain that needed high acidity in blueberry soil.

Once you have picked your spots, space your plants so that they'll have room to grow to fullest size for maximum yields. Set plants about 4 feet apart in rows. If you wish a hedgerow effect, they can be set closer. Remember that if you do this, you'll need to pay closer attention to providing more fertilizer and moisture. When you plant in several rows, leave enough room to walk easily among your bushes at harvest time. A 5-foot-diameter circle usually is sufficient. Blueberry bushes also can be interplanted among your azaleas, rhododendrons, and various pine, fir, or spruce shrubs. So long as they have that acid soil they love, they can be mixed and matched with evergreens to add appeal to your conventional landscape shrubbery.

Blueberries may all look alike in stores, but for your home plantings you have a selection of tasty varieties. Some are ideal for fresh use; others are better for pies, jams, and preserves. Some are multi-purpose. You can get year-old stock, but often nurseries offer two- and three-year-old plants as well. Because blueberries have such a fine root structure, it may be best to select the two-year-old plants. They have a better start and will transplant more readily, especially if you buy them in containers or balled with peat moss in plastic bags.

Most northern varieties are considered self-fruitful. However, it pays to grow several varieties together. If you provide better cross-pollination this way, larger and more numerous berries will result. If you want early, midseason, and late varieties to extend your harvest, be sure to select varieties that overlap in blooming so that pollination is the best possible.

If you have the space and you delight in blueberry pies, plant a selection. Among early types, *Earliblue, Collins,* and *Weymouth* are good. For midseason you can grow *Stanley, Blueray,* or *Bluecrop. Herbert, Burlington,* and *Coville* ripen late.

Earliblue is upright, spreading, and vigorous. It produces medium loose clusters of large, firm blueberries. *Collins* is similar but ripens a bit later.

Weymouth is a smaller, more open and spreading bush. It can fit neatly in border areas around your home grounds. Its quality is not as good as that of the others.

Stanley is erect and vigorous. Fruits are medium loose with medium, light blue, firm, and aromatic berries. They taste great.

Blueray is upright, vigorous, and somewhat spreading. Fruit clusters are small and tight, but berries are very large, light blue, and firm. It is one of the best for dessert use.

Bluecrop is a vigorous bush, upright in growth and nicely productive. Fruit clusters are loose with large, light blue, firm, and aromatic berries, but quality is not so good as Blueray.

Herbert is vigorous, open, and productive among late ripening varieties. It has large to medium berries of good eating quality.

Coville is vigorous, spreading, and productive too. It bears large to medium-size berries, which are firm, tart, and good eating.

There are many other varieties available, from nurseries that serve their own states as well as from those that provide excellent blueberry bushes nationwide.

Rabbit-eye blueberries, the typical southern name for these high-bush beauties unlike their northern relatives, are not very self-fruitful. It is always best to plant several different types to ensure adequate cross-pollination. So it's important that the period of bloom of the varieties overlaps.

In southern areas, attention has been focused on developing improved varieties that fit the climatic conditions of warmer areas. The University of Georgia has been active in developing well-suited new blueberries. Several thousand seedlings have been grown over the past decade. Three show exceptional results. These are *Southland, Briteblue,* and *Delite.*

Southland is moderately vigorous and produces a dense and compact plant, unlike some of its parents, which were too tall for convenient home-garden use. Berries are light blue, medium-large, and firm with good flavor. It blooms when *Tiftblue* does, ensuring better pollination if you plant both varieties together.

Briteblue resulted from a cross made between a native blueberry called *Ethel* and a Georgia variety, *Callaway.* The cross produced an open plant with firm berries that are ready in midseason.

Delite is an upright plant with numerous branches. It bears large round, light blue berries of excellent flavor. Berries do ripen late, compared to other older varieties, and they remain on the plant rather well if you can't pick when they're first ready.

Other good southern varieties include *Callaway, Coastal, Homebell, Tiftblue,* and *Woodard.* Now that blueberries are gaining popularity in many areas for home gardens, alert nurseries are providing a larger range of varieties for your selection.

In fact, experimenters at the University of Minnesota have been working on exceptionally winter-hardy blueberry varieties. They must do so if they hope to develop bushes that will survive in that extreme winter weather. Their efforts have been directed toward crossing low-growing native blueberry bushes of Minnesota with more desirable varieties. The objective is tastier berries on bushes that can withstand temperatures of 25 degrees F below zero and colder. The experimenters have noted that a low-growing habit fits nicely into home-landscape plans and also survives under deep snow cover better than do taller bushes.

If you buy mail-order plants, be certain to emphasize where you live. That may sound unnecessary, but sometimes clerks in the rush of a busy season may not notice that you need warm-weather or cold-weather varieties. Mark your order form clearly so the nursery will make sure that you get the right type for your area. Reliable nurseries often will write back to suggest you change a selection if they believe the one you picked first won't prosper for you in your locale.

Blueberries sold as one-year-old plants normally are rooted cuttings direct from the propagating bed or tray. Their roots may not be well enough developed to withstand any setbacks at planting or during their first year in your land if drought occurs or winter is particularly severe. Two-year-old plants are rooted cuttings that normally have spent a year in the ground at the nursery. They are better prepared for transplanting into a new location. There is little value in buying older plants. They cost more, but maturity for them usually comes not much more rapidly than for two-year-olds that get a firm foothold in your soil.

Best planting time is spring. For each new plant, dig an area about 2 feet in diameter and 6 inches deep. That may sound like a rather large area for such a small bush. However, remember that these are very shallowed-rooted plants. The hairlike roots will spread at or close to the soil surface. That extra pulverizing and preparing of soil when planting benefits them.

Mixing peat moss in to make a surface-soil mixture of one-half peat moss and one-half soil with its own organic content also is a benefit to help plants get the strongest start possible. Begin planting as early in spring as soil can be worked without becoming cloddy or lumpy from excess moisture.

Fall planting is possible with two-year-old or three-year-old plants, but be sure to prepare the soil as well and mulch the plants too. Spread sawdust 4 to 6 inches deep around fall-planted blueberry plants. Peat, straw, or other materials are good too, but sawdust improves and maintains that vital soil acidity. It also conserves moisture, smothers weeds that might rob your plants of nutrients, and prevents excessive freezing and thawing of the ground. Too much freezing and thawing can break roots. This extra fall application of mulch helps insulate the plants during their first winter.

It's a good idea, of course, to maintain continual mulching on blueberries. Each spring and fall, add more. The material you applied early will have decayed somewhat, which is naturally beneficial to the soil's condition. That in turn benefits your blueberry plants.

With all this emphasis on mulching, it may seem as though you are being asked to do extra work. Not so. Mulching, that simple spreading of peat, compost, leaves, or sawdust, takes only a few minutes per plant. It saves you hours of weeding and cultivating and helps you avoid damage to plants. In addition, your plants—and subsequently you—gain the benefits of higher yields.

Whenever you plant blueberries, always keep their fragile root structure moist. Don't tamp the soil as you should for most other plants. Pat it into place, leaving it fairly loose. Those fine roots need all the help they can get initially to penetrate the soil around the bush as they begin the life-sustaining process for your bushes.

PRUNING POINTERS

After planting, remove all fruit buds from the young plants. That way the first year's strength goes into root and branch growth rather than a feeble attempt to set a few berries.

Prune away any dead or damaged wood and short twigs. Fruiting shoots that will bear next year's crop should be about the length of a lead pencil, coming from the branches on one-year-old plants. Round, fat fruit buds will grow on the tip portion of these shoots to bloom next season. Each fruit bud will produce a cluster of berries.

It is important with blueberries that sufficient pruning be done each year to maintain bush vigor and encourage production of a number of strong fruiting shoots each year. Prune with two purposes in mind.

First is to adjust the fruit crop to the capacity of the bush and its root system. Second is to stimulate strong, vigorous shoots to produce next year's crop. If the bush is overloaded with fruit one year, there will be little strength left in the plant to produce enough vegetative growth for future years. Subsequent crops will be less, and the bush will become weak and twiggy with small fruit.

Blueberry bushes send up new shoots each year. These can be

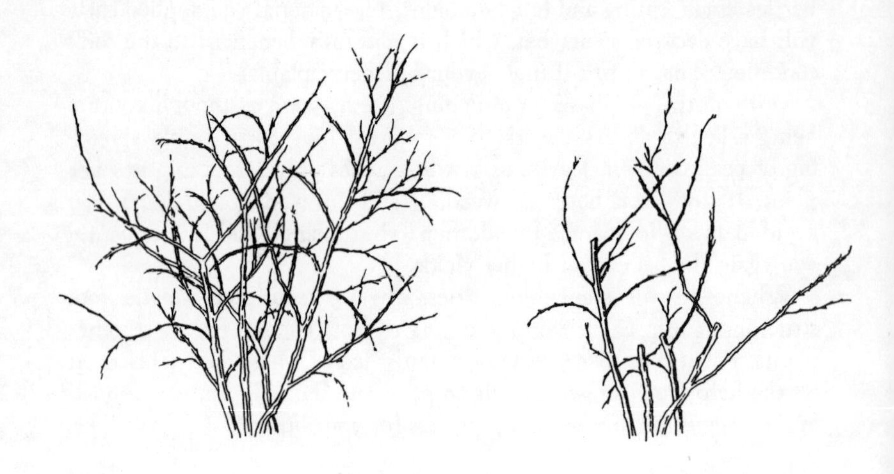

A blueberry bush that looks like the one at the left is sending its energy in too many directions. When the bush is pruned properly, it's like one on the right.

headed back to encourage side laterals that will bear large fruit. Best pruning time is early spring. You can identify fruit buds by their large size and estimate your potential crop. Head back less vigorous, twiggy canes to a strong lateral branch or a new shoot. Remove short twigs and their fruit buds.

Strong vigorous canes can usually be pruned to remove about half of the fruit buds simply by cutting away the weaker new growth. A rule of green thumb for pruning is to leave one fruit bud for every 3 inches of new shoot growth.

Here are some good general rules for blueberry pruning. Limit the number of canes or main branches rising near the crown to one or two for each year of age of the plant. An alternate guide is to limit the canes or main branches from the crown to one or two for each foot of height of the bush. That should leave a maximum of six to ten canes for older, mature bushes.

Remove sucker shoots and all weak, twiggy branches that seem small and spindly. Thin the vigorous fruiting wood to one or two fruit buds per 3 to 4 inches of shoot growth. Keep in mind that each fruit bud will produce six to eight or sometimes more berries.

As your blueberry bushes mature year by year, vigorous new canes will rise from the crown. They may tower over the average height of your bush. Simply head them back to the average size of the plant. Remove older, weaker ones. That way, your blueberry bushes will be automatically replenishing themselves with strong, vigorous new growth over the years. In time, you'll be able to remove some of the older, lower branches. That will let the taller, more vigorous ones with their laterals produce a more convenient, more productive plant in the years ahead. It will also save stooping to pick the berries.

FEED BLUEBERRIES RIGHT

Nitrogen is the main fertilizer element needed by blueberries. Don't add any the first year if you planted one-year-old stocks. Blueberry plants grow in what are known as flushes, one early in the spring—followed by a pause—and another a month or so later. Use a garden fertilizer such as 10-10-10 or 10-6-4, applying about ¼ pound around each 2-year-old plant. The next year, spread ½ pound around each bearing plant. Spread fertilizer on the soil or over the mulch over the entire root zone. That's the area extending from 6 inches out from the plant to the drip line, which is the furthest reach of the branches.

Since nitrogen is the key element, you should get the more readily available forms, especially if sawdust, wood chips, or bark are used as mulch. These cellulose materials also need nitrogen to continue their own decomposition process or they may take it from the soil, reducing the amount available to your plants.

Using ¼ cup of ammonium nitrate around each bush is recommended in spring. When blueberries are mulched, spread ¼ pound of ammonium sulfate, or its equivalent of another ammonium nitrogen carrier, each spring when buds begin to swell.

A second application can be made six or so weeks later. As plants become fully mature and bear from 3 to 4 quarts of berries each, you can increase this amount a bit. After all, when plants take the plant food from the soil, it's up to you to keep them well nourished by resupplying what they remove.

The major pest of blueberries is not an insect. Birds are blueberry lovers, eaters, and destroyers. Fortunately, special netting keeps them off the bushes. Avoid using cheesecloth, since it may shade the berries and prevent them from becoming as plump and sweet as they should be. A few minor insect problems may occur, but consult the chapter on pest prevention for details about any needed sprays.

PICK WHEN TRUE BLUE

Most blueberry varieties will ripen over a period of several weeks. Be prepared to pick your berries several times. They should be fully colored and pop easily from the cluster. Berries usually turn blue several days before they develop their maximum sweetness and flavor. Gently roll them from the cluster with your thumb into the palm of your hand. Those berries not quite ripe usually remain for picking later. With practice you'll get the feel of this simple process.

Avoid overhandling berries because you'll rub the typical "bloom" from the fruit. Its removal does no harm but reduces their eye appeal.

Plant some blueberries this season. In a year or two, with little care other than mulching, pruning, and fertilizing–which amounts to an hour's work per bush per year–you'll have berries aplenty. Bet you can taste those pies already.

14. Grow a Strawberry Shortcake

What's rich and red and ripe and melts in your mouth? Strawberry shortcake does. You can enjoy it much more often when you grow these delicious fruits in your garden. They're the most popular home-garden fruit grown in America. Strawberries require little room or care and are easy to grow and abundant in fruit supply. They perform well in every state. They also are high in vitamin C.

You can grow them in rows, in hills, in window boxes, in patio planters, along a path, or interplanted with flowers and vegetables. They can even be used as a fruitful ground cover near your shrubs. When company comes to call, you can roll out a bright and tasty strawberry barrel to let friends pick their own dessert right from the plants.

For the little space they take, strawberries are the most productive of any home-garden fruit. You can grow them from Florida to Alaska. They have a wide range of climatic adaptation. New everbearing varieties yield extra abundantly and let you pick these treats from spring right up to frost. No home should be without them.

With proper, easy-to-provide care, just twenty-five plants in less than 50 feet of row will yield 25 to 40 quarts of berries. That's bountiful abundance. By selecting several different varieties, you can enjoy strawberries from spring right through the summer into late fall.

Although the third year is the most productive, strawberry plants have a way of replenishing themselves. Runners from mature plants produce new plants that will produce crops in future years. You might say that strawberries just naturally recycle themselves every few years. That's helpful. Once you set a bed or two, they'll give you not just their tempting berries but also new plants to continue growing more for many years to come.

To determine the best for fresh, frozen, or preserving use, check the listings in those handy nursery catalogs. Because strawberries are so popular, most every mail-order seed company includes the better varieties of this fruit as well. Here are some of the best ones, developed carefully by plant breeders to produce the highest yields of the most flavorful, hardiest, and most season-stretching berries. Old-time disease problems have been virtually eliminated by the use of virus-free plants and improved varieties resistant to problems that once caused poor crops or killed plants. Today, strawberries are probably the easiest fruit crop to grow, especially on limited space.

Insist on virus-free stock. Nurseries have it and can guarantee that this parent stock is free from injurious disease problems. The slight additional cost for this stock is minimal compared to buying non-certified stock that may give you problems.

Your choice is wide. Since 1920 more than one million different seedling varieties have been evaluated by U.S. Department of Agriculture researchers and experiment-station workers in various states. The best have been introduced to fit the needs of various growing regions. It pays to compare test plantings of new offerings from time to time, since improved varieties can outyield even old favorites.

Sunrise is an early, large, tart, and bright-colored fruit.

Raritan is an early midseason berry of excellent quality. It is very productive with high quality, attractive fruit.

Surecrop is another midseason one with large, light red, firm, and tart berries. It is good for freezing.

Sparkle is midseason, very productive, medium-firm, high-quality, and vigorous in producing new runners for new plants.

Catskill is midseason and very productive and has large berries. Fruits tend to be soft and are best used fresh.

Jerseybell is late, productive, large, and of good quality.

Fletcher is vigorous and produces many runners. Flowers seem somewhat resistant to frost. Berries are medium, firm, glossy, and of good quality. It is a fine variety for fresh use or freezing, slightly better than Sparkle.

Gala is very early with large, slightly dark and rough berries. Plants are vigorous and productive and fruit well without crowding. Avoid frosty areas, since this variety blooms exceptionally early.

Garnet is a promising new variety. Berries are large, medium red, moderately firm, and of good quality.

Holiday is a newer, firm-fruited variety. Plants are productive and vigorous and make well-matted rows, fruits are large, very firm, bright red, and glossy. They're aromatic too and ripen midseason. Holiday is fine for fresh or freezing use as well as jam.

For easy reference when you shop, the following are all-round recommended varieties.

For early season, try *Earlidawn* and *Sunrise.*

For midseason, try *Midway* and *Surecrop.* For late season, try *Sparkle.*

Among the newer everbearing strawberries, *Geneva* is a fine one. *Ozark Beauty* also is delightfully sweet. *Gem, Superfection,* and *Streamliner* are others that do well.

Because nurseries are continually introducing even more improved types, check their recommendations. However, exaggerated claims you see in the mass magazines for plants that do extraordinary things should be viewed with caution. As gardening has become more popular, some mail-order firms have sprung up that seek to exploit this increased interest in gardening. What they say in full-page glowing ads may be far from the whole truth. Besides, some of these firms have never even been in the plant business before.

When you shop, either locally or by mail, depend on well-established, reliable plant nurseries. The list at the end of this book includes those firms that have been reputable for many years.

GOOD SITE GUIDE

Although strawberries are available to thrive in any climate and do have a fair tolerance for various types of soil, they perform most prolifically when the site is right and soil is to their liking. Both factors can be provided.

Pick an area with lots of sun. Avoid frost pockets to prevent nipping buds when they bloom in early spring. A gentle slope is perfect; so is an area protected from harsh winds. You can and should mulch over beds in cold climates to protect the plants, but that's a simple job each fall.

Strawberries prefer soils with high organic matter. If your soil lacks it, you can easily till under compost, peat moss, manure, or similar materials to enrich the existing soil.

If your space is really limited, you can create a strawberry pyramid and fill it with improved, ideal soil mixture. You say you have less space than even that requires? Well, try a window box, a patio planter, or even a fascinating and productive strawberry barrel. There's no good reason you can't find some place for these sweet berries, even perhaps a hanging basket or two on a balcony high above a city street.

Where possible, plant a "green manure" crop of rye the previous year. Whether you do this or not, plan to apply 2 pounds of a high-phosphorus and high-potash fertilizer such as 6-24-24 per 100 square feet of area before turning the soil. Nitrogen, as you know, is preferred to build vegetative growth for trees and bushes. For strawberries, to attain their best fruit set and sweetest taste, you must rely on phosphorus and potash more. That's their function as plant foods, building fruit set, good roots, and heavier yields of crops.

Work soil as early in the spring as possible—as soon as it is warm and crumbly. Strawberries should be planted early. Avoid using ground that has been planted to tomatoes, peppers, eggplants, or raspberries within the last two years. Some diseases that afflict these plants may remain in the soil to slow or stunt strawberry plants. Be sure that your soil has good drainage. Strawberry roots are shallow but, like those of all plants, don't like to be wet. When air pores in

Strawberries need to be planted at the right depth. These plants, from left to right, are planted too deep, too shallow, and just right.

soil are constantly clogged with water, the roots can't pull up the needed nutrients or reach out to feed as well.

SPACING STRAWBERRIES

You can select from three types of culture:

1. The matted row, in which plants mass together.
2. The spaced row, in which you remove some plants to give those remaining more room to spread individually.
3. The hill system, in which individual plants are given ample room to perform to perfection by themselves.

Each type of culture has its merits.

For matted rows, space plants 2 feet apart. Most commercial growers prefer the matted row method, since it produces greater yields per acre. That's not so critical in home plantings, since you can get fine results with the other methods. If you have the room, plant several rows 4 to 5 feet apart. Runners (so-called daughter plants) are sent out by the parents to root in all directions. Your rows will eventually become densely populated by new plants. Cultivation is somewhat difficult with these tangled matted rows. Plants are matted; they require little weeding or other care but produce lots of berries. Final width of the row should be about 2 to 2½ feet. In matted rows, plants will be about 8 inches apart when the row is filled in fully. The advantages of this system are high yields and ease of care for harvesting. Disadvantages are crowding of plants, slightly smaller berries, and susceptibility to disease and drought.

The spaced row system begins the same as a matted row. Space plants 2 feet apart with rows 4 to 5 feet apart. The object is to have a bed 15 to 18 inches wide, with new plants from runners spaced 7 to 10 inches apart. You should remove the excess plants that are too close together. When you remove the extras, the nutrients, moisture, and sunlight are concentrated on fewer plants and encourage stronger growth and bigger berries. You can modify this spaced row system even further by removing more plants.

With the hill system, you should remove all runners as they form. That way, original plants are the only ones that produce a crop. Distance between rows can be narrow, since rows are really only one plant wide. Spacing within rows should be close, about one plant every foot. You can elect to plant several parallel rows if you plan to remove all runners. When you remove runners, individual plants develop

Strawberries are best grown in hills (top) or rows (above). The row may become either a matted row or a spaced row, depending on how you handle it.

large and numerous crowns. Because plants get large, they'll bear more fruit than typical individual plants in other systems.

The advantages of this hill method are larger and more berries per plant and easier weeding and picking. However, you'll need more plants to start your strawberry patch. Also, it does take extra effort to remove those constantly sprouting runners every few weeks during the spring and summer. Another point is worth considering: without letting plants set new runners, you'll have no new plants to start new areas or replenish the bed as older plants become less productive.

Strawberry plants are most prolific their second and third years. After that, the original plants become tired. They just don't bear well in future years. It's important to let runners multiply to some extent so that your beds replenish themselves. For this reason, a modified matted row seems best for home-garden purposes.

TRY A STRAWBERRY PYRAMID

Here's a handy way to grow strawberries even when ground space is limited. Build a strawberry pyramid. All you need is an area 6 feet by 6 feet. You might consider this a postage-stamp garden, but it works wonders for abundant harvest.

Make the bottom level 6 feet square, with 2 × 6-inch planks. Make the second level 4 feet square and the third (top) level 2 feet square. Redwood boards are best, since they resist decay in contact with soil. Lay the largest frame on a level surface in a sunny location. Fill it with soil first.

Prepare your soil mixture using 13 bushels of topsoil, 5 bushels (or the typical 6-cubic-foot bale) of peat moss, and 5 bushels of perlite or sand. Mix one pound of 5-10-10 or 6-12-12 fertilizer with the soil. You can blend the material with a shovel, mixing alternate layers until the consistency is even. Then fill the first frame, firming the mixture to avoid later settling.

Next, position the 4-foot-square frame. Fill it and firm the soil mixture. Then add the final frame and fill that. Tamp soil down firmly. You may anchor the frame with stakes in the ground if you wish. Keep the soil weed-free until you're ready to plant.

By setting plants 9 inches apart, you have twenty-eight on the first level, sixteen on the second, and nine on the top. This gives a total of fifty-three plants in the 6-foot-square pyramid. The top level can be somewhat more closely spaced.

Plan to water the pyramid each week, especially when plants are setting fruit. You can mulch all three levels with peat moss to help retain soil moisture. A pyramid this size lets you enjoy a new dimension in fruit growing, weed and tend it easily, and reach across from any side to pick your plump, ripe harvests.

A variation can be decorative. Some gardeners set a dwarf fruit tree in the top, digging down far enough for its roots to be well spread. With this idea, you'll need to pay more attention to feeding and water-

PYRAMID STRAWBERRY BED

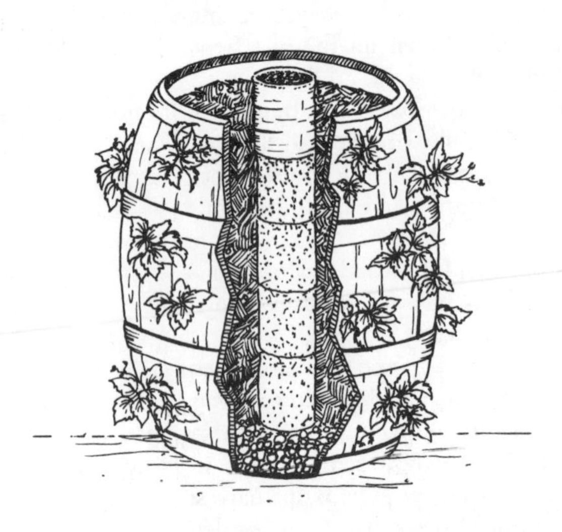

If lack of space is a problem, you can grow strawberries in a square pyramid (two views, top) or even in a barrel (above).

ing each year to supply the nutrients a dwarf tree and your strawberries need together. However, such an arrangement does provide an interesting and multipurpose planter for fruitful gardening, even where your space is tight.

ROLL OUT A BARREL

If you have even less space, say on a balcony or porch, you still can enjoy luscious ripe strawberries in season. Try a strawberry barrel. Hardware stores and garden centers now offer these wood barrels or variations made from ceramic material. Or you can make your own from an old nail keg or storage barrel from a flea market or garage sale. Never, however, use a barrel that has contained chemicals or other materials harmful to plants.

The first step is to drill holes around the barrel as shown in the illustration. They should be about 1½ inches in diameter. Drill them 6 to 8 inches apart, depending on the size of your barrel. You can start by painting or whitewashing the barrel if you'd like to match the color of your house or patio furniture.

When the barrel is dry, set it on the ground or raised on four bricks. Fill the bottom with 2 to 4 inches of coarse gravel, small stones, or broken flower-pot parts. This layer provides good drainage. Excess water can escape rather than clog the soil and rot roots. From a plumbing-supply store, buy a piece of perforated pipe (or you can make your own from stovepipe by punching holes in it). Place this in the center of the barrel supported by the gravel in the base.

Next step: fill the barrel bottom with prepared, enriched soil mixture of the kind I just recommended for the strawberry pyramid. You can use leaf mold, peat, or compost as you combine these organic materials with good topsoil. If you live in an apartment and have no access to good garden soil, remember that garden centers often sell smaller bags of peat moss and dried compost as well as 25-pound bags of potting soil. Fill the barrel to the level of the lowest circle of holes. Then place your first tier of strawberry plants through these holes. Spread roots horizontally and evenly and add more soil. Be sure crowns of plants are just at the opening in the barrel. Continue filling the barrel with soil and planting until you reach the top. Then pour gravel into the pipe. That gravel core provides a fine watering system and also lets excess water drain back from the soil after a rain or in case you overwater.

You can plant twelve to twenty-four strawberries in a small-to-medium barrel. The larger barrels will accommodate even more. Care is similar to that for ground-planted strawberries. A little balanced fertilizer, stronger on the phosphorus and potash sides of the formulation, can be mixed with water to feed your barrel plants during the season as they begin to set their fruits.

Strawberries also make a lovely hanging basket. Just a few in good soil will reward you with a pint or more of berries. One neighbor of ours planted a dozen strawberries in hanging pots all along her porch. They flowered nicely as a plant display and produced several pots of berries. Window boxes are another likely spot for strawberries. So are patio planters.

Since most of you will have the room to plant strawberries outdoors, here are the steps to follow. Don't let the roots dry out. Nurseries normally ship bare-root plants wrapped in moist sphagnum moss. Garden centers sell plants in polybags or composition paper pots. If the ones you buy are bare-rooted, put them in a pailful of water containing one cup of compost. Plant the strawberries as soon as you can.

Strawberry planting requires attention to that just-right level of the crowns to ensure success—not too high, not too low, but with the crown just at the soil surface. This requirement may seem picky, but your plants will know the difference if you don't set them at that just-right height.

Although your mouth may be watering for strawberries the first year, resist the urge to let them bloom and set fruit. That's fine with everbearing varieties that will bear a crop in the fall. But all spring- and summer-bearing varieties need the first year to become established. It pays in terms of future abundant harvests to pinch off flowers that appear the first year, except on those everbearing ones. Pinch you should, and again whenever flowers form. That simple practice will encourage ample runner formation, which is your primary objective the first year, unless you prefer the hill system of individual plants. Those new plants formed at the end of runners will produce the major share of your crop next year.

If plant growth seems slow in mid-June, apply 1 pound of a complete-analysis 10-10-10 or 12-12-12 fertilizer per 50 feet of row. Spread ½ pound along each side of the row. Scratching lightly and watering after application lets the plant food dissolve and go to work to boost strawberry growth. Keep fertilizer particles off the leaves or you'll see some burning from chemical fertilizer.

Weed you must to let your strawberries get the most from their

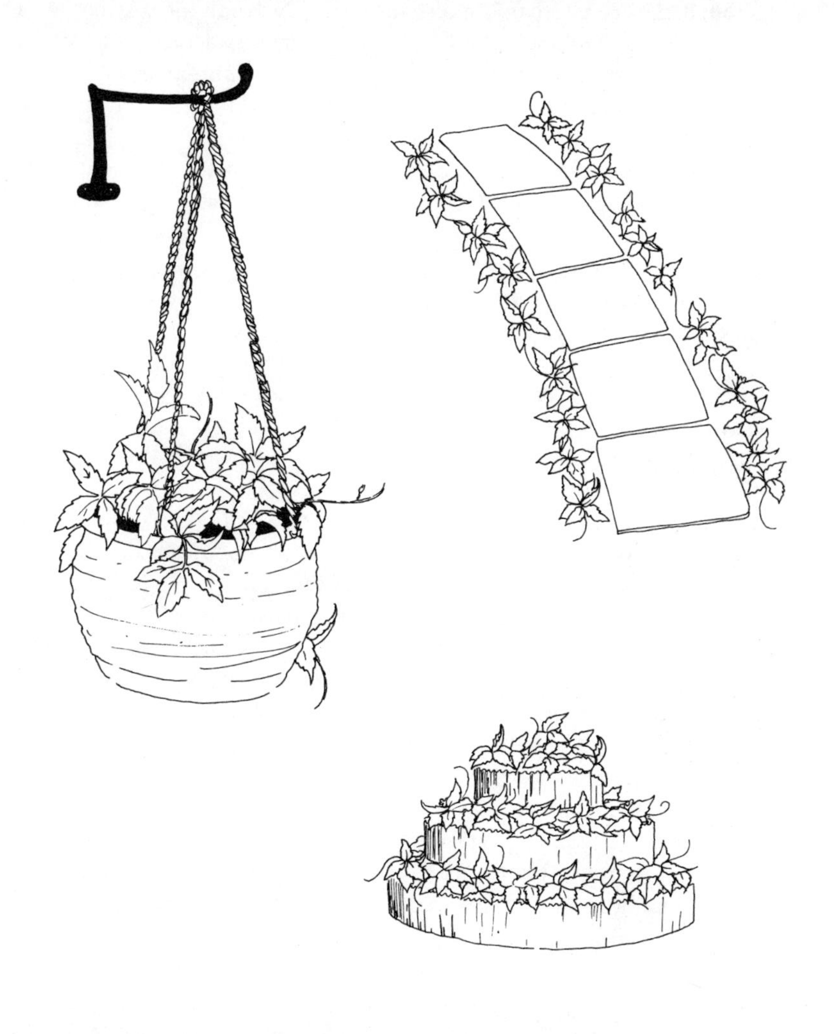

Other ways to grow strawberries are in a hanging basket, along the edges of a sidewalk, or in a round pyramid formed from metal edging for flower beds.

new home. Weeds steal nutrients and moisture. Pull larger weeds and mulch as much as possible with organic materials to smother smaller weeds. Another possibility is to place black plastic in position before you plant. Then simply slit holes and place strawberries through the slits. Organic mulch is better. Not only does it conserve moisture, smother weeds, and look nicer, but as it decays each year it adds some nutrients to the soil and improves its texture too.

By the second year, you'll be enjoying more fruit than you thought possible from so few plants. Strawberries are tastily prolific. The third year you'll also find that your crop is heavy. After that, you may notice a decline in yield. Generally three years is as long as a strawberry bed is at its most productive.

At that time you have some options:

1. You can remove the older, tired plants and let the younger ones that sprout from runners become the parent plants. This simple renovation takes a few hours but is well worthwhile.

2. You can also elect to transplant the younger new plants and start another bed.

3. We have found that rotating several areas every few years works well.

First you start a bed. By the second year, as you remove the excess runners to prevent an overly matted row, begin another bed. After the third year, when the initial bed begins decreasing in productivity, the next bed is increasing. The fourth year, remove the original bed and renovate the soil by adding more organic matter. Then replant it with excess runner plants from the second bed. In this way, you merely rotate your renovation plan to keep those berries as rewardingly prolific as possible.

In areas with severe winters, you may need extra protection to prevent frost heaving that can uproot and kill strawberry plants. Old bales of hay or straw work well. So do piles of dried leaves from around your trees. However, if any trees drop diseased leaves, especially from fruit trees, don't use them. No sense risking infection of your strawberry bed.

INSECTS CAN BE STOPPED

There are some insects that enjoy strawberries as much as you do. Leafhoppers, aphids, earwigs, slugs, and spittlebugs are some of them. Consult the pest-prevention chapter for information on how to beat them back and preserve the bounty you deserve from your strawberry beds. Slugs can be controlled by catching them in saucers or pot-pie tins filled with beer, or by using slug bait.

Each spring as the weather warms, remove the cover mulch you may have applied to protect your plants during cold winters. Remove it in two phases: about half as snow is gone, the rest from the top of plants when spring is really warm. Leave mulch around the plants for smothering weeds.

If you planted early-, midseason-, and late-ripening types, strawberries will begin bearing in June and last late into summer. When to pick is easy to determine. As berries become red and plump and ripe, just pick a few. Since each plant bears its own abundance, several pickings will be necessary. Your taste and the berries' appearance are the best indicators of harvest time.

If you have any surplus, you can save it. Strawberries are easy to freeze, which is the best way to save them for use throughout the off-season. Wash them first, then pack them fresh in quart or pint containers. You might prepare a syrup as described in your favorite cookbook. Jam is another superb way to preserve the bounty from those strawberry beds, borders, tubs, or planters.

Pick a spot right now. Begin the soil preparation. Strawberries are America's favorite fruit. They'll taste good home-grown.

15. Enjoy Nuttier Living

Be nice to nuts and they can reward you well. Actually, to be accurate, nuts are fruits. They just happen to be the fruits of nut trees. That's understandable.

Although initially I did not plan to include nuts in this book, I changed my mind one sunny day as the editor who helped organize it sat with me shelling a few black walnuts that had dropped from a native old tree on my back garden area.

"They are mighty hard to crack," he noted, "but once you learn the knack of picking them, there's nothing like black-walnut cake. Why not add nuts to the book?" he asked me.

I agreed. After the fact, it seemed quite logical to include some of the more easily grown and tastier nuts. Nut trees, once established and nursed through their traditionally slow start, are about as carefree as a tree can be. There's little pruning needed and nearly no other care. Fact is, there's lots to like about nuts.

Edible nuts are important horticultural crops in many sections of the United States. The value of almonds, filberts, pecans, and walnuts is well over $60 million annually.

Commercial growing is centered in the best climatic areas, where soils also are the most favorable. That's in the South, generally speaking. Pecans do well across the southern states, excluding Florida. They thrive in Georgia, Texas, and New Mexico, where the tree is a native. Almonds and English walnuts are grown widely as commercial crops in California. Filberts are favored in Oregon and Washington.

However, nut trees do grow surprisingly well outside these most favorable growing areas. By choosing the right types and varieties, you can grow some nut trees in almost any state and in parts of Canada as well. In addition to being durable sources of shade, nut trees have that extra added attraction every fall—a tasty, richly rewarding harvest of nuts.

Nut trees are larger than most fruit trees. They require a lot of room. Some nuts are produced on bushes that fit better in landscape plants.

Some of the most popular nuts for home gardens: top—hickory and pecan; middle—almond and Chinese chestnut; bottom—hazelnut (filbert) and walnut.

Nut trees typically have taproots. That means you must dig in deeper and better to get them growing well. However, once that initial work is done, they can be remarkably hardy and long-lived.

Heres a brief checklist of the most popular nut trees that can be grown and in which areas. They've been graded by a leading nursery, Henry Fields Seed and Nursery Company of Shenandoah, Iowa, a firm that has been highly experienced with fruits and nuts since it was founded back in 1892.

Butternut trees yield rich, buttery-flavored nuts. America is divided into ten horticultural zones based on lowest mean average temperatures. Trees are usually hardy in horticultural zones 8 through 4. Trees may reach 40 to 60 feet tall. They are best planted in spring, about 30 feet apart. These tasty nut trees prefer rich soil, but otherwise no special care.

Chinese chestnuts have replaced our American chestnut trees. A blight some years ago swept the nation, killing practically all our native chestnuts. Attempts have been and are being made to bring back specially produced strains, but only time will tell whether these valiant efforts will succeed.

Chinese chestnuts are hardy from zone 8 through 4. These trees appreciate well-drained soil. They should be spaced 20 feet apart and prefer spring planting. Trees grow rapidly and bear early to provide you with large, sweet nuts. The trees themselves resemble in shape, form, and size a standard apple tree not yet at its prime maturity.

Hazelnuts (filberts) are another hardy landscape addition. The nuts are produced on vigorous shrubs that mature to about 6 feet tall. They are hardy in zones 8 through 2, which includes just about all but the most southerly and northerly portions of our country. As shrubs, they reward you with red and yellow foliage in fall after you have harvested the nuts in late summer. Hazelnuts prefer well-drained, rich soil and should be spaced 10 to 15 feet apart. They produce abundantly and nuts are easily cracked, unlike some others you might grow.

Hickory trees are native too. These tall, hardy, pyramid-shaped trees prefer rich, well-drained soil. They are hardy in horticultural zones 8 through 4. If you keep them heavily mulched until they become established, they'll bear year after fruitful year. Hickory trees may reach 50 or more feet in the sky, so allow them ample growing room.

Pecans are more restricted in their habitat. They are hardy in zones 9 through 5, which does limit them to southern states. Pecans are large, spreading trees. They'll need 40 to 60 feet, measured in diameter, to reach their full potential. They prefer a deep soil with a steady moisture supply. Hardy pecans grow vigorously. They are handsome ornamentals and shade trees well worth a spot in your plantscape for their attractive growth habit as well as their nuts. Pecans do respond to special cultivation and with this extra attention reward you more bountifully.

Walnuts are nicely rounded trees with deep green leaves. They are hardy from zone 8 through 4 and prefer rich loam where they can set strong roots. New varieties are nicely thin-shelled, which is a help at eating time. Plant them in the spring some 50 feet from each other or other trees.

Filberts grow on a smaller nut-bearing bush or tree. They may reach 15 feet tall as trees, or only 8 feet high as bushes, depending on the type you buy. You can count on them for greater hardiness in northern areas. Filberts actually do best in colder areas rather than the South. To ensure a crop, you must plant at least two for cross-pollination.

Other trees also have this unfortunate pollination problem, some more than others. Check catalog listings so you can plan for two where cross-pollination by another variety of the same tree is required to produce the needed fruit set. Most nut trees will require a companion for this purpose.

SELECT THE RIGHT SITE

Nut trees will reward you for many years, but they do try your patience with their notorious slowness to become established. They may seem to linger at the same size for several years before they take a firm roothold. Since most larger nut trees are mainly taprooted, rather than blessed with spreading roots as most fruit trees are, you should pay particular attention to the proper site. Pick an area that has the deepest soil possible. It should be well drained, have a rich and loamy type of soil, and receive abundant sun to help the tree grow to its maximum.

Nurseries sell young saplings as well as older balled and burlapped nut trees. Younger ones, although slow to get started, are the best buy. Before your tree arrives, dig deeply in the spot you've selected. Remember, taproots shouldn't be bent. They may look like a long, skinny parsnip, but from that root will grow your favored tree for many years to come. It deserves the best home you can provide. Dig down 3 feet if possible. If subsoil is poor below the topsoil, remove it. Prepare a mixture of compost, mixed one-half with peat. Mix that half and half with topsoil.

As soon as your tree arrives, plant it. Set the root as deep as it needs to be. Add your improved soil mixture. Tamp it every few inches that you add. When the hole is half filled, soak the soil. Then add more soil until the hole is full. Soak again. Leave a saucer-shaped depression as you would with a fruit tree. It is even more important to direct water to that taproot, since it doesn't have a broad reach to go in search of moisture as easily as other trees.

If the nut tree is small, stake it to give it the support it needs the first few years on your land. Use rubber hose parts to protect the bark if you use wire to tie the tree to its stake. Stout cord also may be used and shielded with old hose sections. You can also buy tree-staking kits with all the parts to keep the tree from bending in wind or during winter storms.

The next step may be the hardest if you're the sort who can't bear to take those pruning shears to growing plants and trees. But prune you must to help that tree get off to the best start possible. Remove at least one-third to one-half of the top. That's right. It may sound severe but this is really to your tree's advantage. This pruning compensates for loss of root hairs and small rootlets from the main root when shipped and transplanted. It also encourages that all-important new growth of your tree, the side branches that become the framework of your mature nut-bearing tree to be. By *not* pruning you'd risk formation of many small, weak sprouts that compete with each other. That first pruning, harsh as it may seem, is vitally necessary.

Mulching is my personal preference. A mulch around new trees preserves soil moisture that these taprooted nut trees need in order to start vigorous new growth. Mulch also reduces your temptation to mow grass too close to a tree planted on your lawn. By avoiding that problem, you won't bruise tender young bark.

Feeding young nut trees is a wise practice. You can mix liquid fertilizer solutions in water or use fertilizer sticks that you hammer into the ground around the tree. These slowly release plant nutrients to feed the tree as it needs more nourishment, month by month.

Another handy tool is the Ross Root Feeder. It attaches to the end of a garden hose. Just drop a few tree-feeding pellets in its container, turn on the faucet, and let the water dissolve the nutrients.

They are carried into the soil with the water through the hollow spike that you easily insert in several spots around the trees.

Once your trees set their roots and begin feeding in their new home, your job is almost over. Check for any narrow crotches, since these can split in winter storms. There's little need to worry about heavy crops weighing down the branches. These hardwoods can take abundant yields with little worry.

Do keep a lookout for damaged branches and limbs on older trees. Judicious, proper pruning will deny decay or insects a point of entry.

Since *filberts* are so handy in many northern areas as a bush for home landscape use, you can shop for several improved varieties.

Italian red has a large pointed nut with medium-thick shell. This tree is vigorous and one of the most productive of filberts.

Potomac is a vigorous hybrid between American and European varieties. It is hardy with attractive, reddish brown nuts, somewhat rounded in shape.

Walnuts, both black and hardy English types, are suitable for many areas too. The hardy English strains are often called *Carpathian* walnuts. They have superior tolerance to midwestern and northern winters. Regular English or Persian types may freeze back or be winter-killed.

Colby is a grafted variety of medium nut size. It cracks handily. Kernels are flavorful. Trees usually begin to bear in five to eight years and are very hardy.

Lake is another grafted variety that bears large, flavorful nuts. This tree, too, is hardy.

Among *black walnuts* that are native Americans, you'll find at nurseries same varieties adaptable to southern areas as well as northern climates.

Thomas is a grafted variety that bears early, producing good yields of reasonably thin-shelled nuts. It may be somewhat susceptible to leaf diseases.

Meyer is a grafted one that bears early and produces fairly large yields of thin-shelled nuts. Nurseries do offer seedling trees produced from outstanding native ones that produce satisfactory nuts. Yields and quality, however, usually are inferior to grafted varieties.

Most nut trees grow quite well as home-ground specimens or shade trees. Pecans, however, respond to clean cultivation around them and fertilization for greater yields. Since soil conditions and varieties vary somewhat, it is best to check with your county agricultural agent for exact recommendations for your area.

With the increase in concern for our environment and the trend to return to more natural ways of life, more families have found that wild nut trees let them do their natural thing. It's true that with care, you can transplant some wildlings to your home grounds. Or you may find a tree or bush with nuts on the land when you move in. Many older homes were owned by people who saved the original nut trees they found when they built or bought the home.

Here's a checklist of some native trees that you may find. If they aren't growing on your own land, always ask permission to transplant any.

The *black walnut* is probably the best-known native American nut tree. It has hard, rough-textured bark and a tall, spreading growth pattern. Its round nuts are encased in a tough husk inside a softer greenish hull. Nuts drop in the fall. They should be hulled as soon as the soft outer hull can be dented with your thumb. Wear gloves. The stain from the hull can discolor your hands. It also can discolor the nutmeats and give them an off flavor if you don't hull them on time.

After hulling, wash nuts with water and dry them on layers of absorbent paper. Place them in a shaded, cool, dry place with good air circulation for several weeks.

One way to remove tough, dark husks is to run over them with an automobile on a hard surface driveway. That may sound strange, but it works. We've done it that way for years. A hammer is handy too, but nuts tend to shatter.

Hickory nuts fall from those shaggy-barked hickory trees, called, not surprisingly, Shagbark hickory trees as well as other hickory trees. The small-to-medium-size nuts are encased in hard, thin husks. These can be cracked with nut crackers without too much effort. Gather them as they fall, and dry them on old screens. After a few weeks of drying, open a few to determine when the kernels inside are crisp. That's the time to store them. Store nuts in a mesh bag in a cool area with good air circulation.

Beechnut trees have smooth, light gray bark. Leaves turn brilliant yellow in the fall. Nuts are small and have sweet meat. The hard kernel is surrounded by a somewhat prickly husk. These, too, should be dried for a week or so on old screens. You can crack them as you would the black walnuts. They are easier to crack if you soak them overnight in water.

If you prefer to crack and pick out the nutmeat from wild nuts, be sure to use it soon or dry it slightly on cookie sheets in a warm oven.

Then, to store it for a few weeks longer, place it in tightly covered jars in the refrigerator or freezer. That helps prevent nut oils from becoming rancid.

Whether you plant domesticated nuts or gather some from the wild, remember that squirrels are just as quick to spot the time of harvest as you are. Keep an eye open for those four-footed nut eaters. They'll squirrel away a large part of your harvest if you don't get to picking first.

Enjoy nuttier living from your home plantscape. Nut trees have that extra added attraction every fall, their nutty harvest.

16. Space Sculpturing Is Tasteful Too

For centuries the horticultural practice of espalier (pronounced es-pal-yay) has had a well-rooted tradition in France and elsewhere in Europe. It can be a most satisfying, creative, and compliment-earning way to use trees and shrubs around your home grounds.

Results of espalier culture are dramatic. They can win you cheers from friends and neighbors as well as the real-estate broker when you may wish to sell your home. After all, the better your landscape, the more attractively it sets off your home for prospective buyers. Realtors have long known that well-landscaped homes with lovely specimens of trees and shrubs can greatly increase the value of property. That's a proven fact and well worth noting. Fact is, a tidy berry patch, some attractive fruit trees, and a landscape that looks as though the owner cares has an important psychological impact. When grounds are well tended, prospective buyers seem to sense that the house itself is in good repair and perhaps worth more than a similar home that has a cluttered, poorly landscaped look.

Espalier can achieve dramatic effects and be a daily pleasure for you as well. In addition, it offers an enjoyable challenge. By close pruning in order to control growth, you train the trunk and branches to lie flat in one place. The practice was widely used by European gardeners in the Old World to conserve space in small orchards and on their home grounds. Today the practice is making a striking come-back as a way to introduce decorative accents or conversation-piece plantings to home plantscapes.

An espalier is simply a living sculpture in your garden. It can brighten walls, fences, proper borders. A beautifully and cleverly trained plant is especially effective against a blank wall where close paving with driveway, path, or street prevents the use of a fully formed and naturally shaped plant.

Fruits trained in an espalier design, such as these apples on wires, can produce a spectacular effect in areas of odd dimensions.

In cities, driveways often take up a large share of the front yard, especially between homes on narrow-frontage lots. In such an area, one large espalier tree can dramatically accent a garage or yield a fascinating display against a neighboring wall that would otherwise be an eyesore. One plant also is much less expensive than a large planting. And despite its necessary periodic precise pruning, it does require less general care.

One caution is in order. The unusual effects of plants trained with espalier methods are attention-getters. Too many espalier plants, however, may detract from the total appearance and give an over ornate, cluttered appearance to the scene. As you consider espalier, use the same judgment as you would in selecting specimen trees and shrubs. They should have enough room to provide the show for which they are intended without undue distractions.

FIRST STEPS

It is usually best to begin your espalier efforts with a young, untrained tree. You can buy plants that have been given their first pruning for the espalier look in the nursery or at the garden center. There is, of course, usually a premium price on these prestarted specimens. It is just as easy to select a likely tree that has a sturdy young trunk and well-formed, nicely balanced branches that will respond to your careful pruning. Most nurseries and garden centers have a reasonable selection of apple, pear, peach, and other suitable fruit trees as well as grapevines which lend themselves to successful espalier cultivation. Look for specimens that have sturdy, well-balanced limbs off the main trunk. Some shrubs and trees branch near the ground. They may not be suitable for regular use, but they can be just the right ones for your espalier plans.

Some trees that have been crowded in nurseries also may have somewhat flattened growth, with limbs already tending naturally in two directions and poor limbs elsewhere that can be removed to accentuate the flattened growth pattern. That's finc. If you ask for shrubs and trees for espalier use, you may find that an extra price is required. However, if you know what to look for, and comment that this flattened, poorer-looking tree might respond with your care, it may be yours for a lower price. Obviously a flattened tree isn't the most desirable for use on a lawn or in a conventional landscape setting. Both you and the nursery win on such a deal.

Although branching may not be perfect on the young plant, you should insist on a healthy, sturdy tree or shrub. Some are grown right in the nursery. Others are balled and burlapped, while others are container-grown. Be certain, as you would with trees for other parts of your grounds, that the bark is not damaged, that there are no broken key main branches, and that there are no signs of insect or disease damage. It is possible to bring poor trees and shrubs back to health, but you're wiser to start off with a healthy one. Why handicap yourself?

BACKGROUND CONSIDERATIONS

Many materials can be used in developing an espalier support system. They can range from heavy-duty wire to wood poles, from the reinforcing steel rods used in building construction to more elaborate trellises. Keep in mind that the design of your shrub or tree and its foliage is what you want to see, not the supporting materials. As the young tree is being trained, the wires, poles, or other supports will be visible. As foliage fills in, even those supports that must be retained will be less obtrusive. You can, of course, paint supports to blend with the color of the wall or side of the background building. You also can paint them green, so they tend to disappear in the foliage of your living plants.

FRUITFUL ESPALIERS

Espalier fruit trees provide a tasty extra added attraction in your fruitful landscape. Especially when growing room is scarce, fruit trees and grapevines perform exceptionally well in espalier form.

You should select fruit trees grafted onto naturally dwarfing rootstock. They are designed for more unusual uses and won't range out of proportion. Semidwarf trees may be suitable for those large expanses, including hedgerow growing as property borders. However, dwarfing rootstock is probably your best bet.

You can choose whichever fruit you like best. Apple, pear, peach, quince, nectarine, crab apple, and even fig can be well trained into classic espalier shapes. Among these types you can usually find a number of favorite varieties.

With apples, for example, we have used dwarf varieties of *Northern*

Spy, McIntosh, Yellow and Red Delicious, and *Lodi.* Local sources may offer different varieties that fit better into your climatic conditions. And nowadays there are more dwarf pear trees available than were available years ago.

For truly limited space, dwarf peach trees are a delight. They grow only a few feet tall, but they can be shaped to delightful patterns.

THE SHAPES TO COME

When it comes to creating espalier designs, let your imagination wander. With care, you can create the classic branched candelabra with its upright towering limbs. Depending on the space you have, this can have three, four, five, or more branches. One friend even grew a menorah tree in honor of his Jewish faith. It became the hit of the year among the gardening friends of his temple.

You might try a checkerboard design or even a herringbone. For apples and even more appropriately for grapes (since it suits their natural growth habit), try a cordon espalier. The tic-tac-toe design is nicely distinctive as you train the trees to form a neat pattern of blocks and squares. This can be changed by proper pruning into a triangular pattern, more like the checkerboard style.

For an even more exotic shape, the Swedish Christmas tree has eye appeal to many. This type of espalier is more easily accomplished with vines and plants such as pyracantha than fruit trees. Yet, with patience and judicious pruning, you can create such highly personalized effects with dwarf fruit trees.

The basic patterns shown in this chapter give you some idea of how to begin this periodic pruning in order to achieve espalier displays. As trees grow year by year, the simple pruning needed to maintain their shapes takes little time. A snipping in the fall when leaves are gone will realign the basic shape. Then summer pruning will keep side-branching under control.

Some gardeners believe that espalier pruning will cut into their fruit crop. It may, but not quite so much as you might imagine. Remember that fruit trees naturally respond to pruning with a desire to compensate for the growth you cut away. They'll surprise you with the abundance of bloom they set on new shoots. Remember, too, that by opening plants to sun and air, and reducing the number of fruits formed, you'll be rewarded by larger, healthier, easier-to-harvest crops.

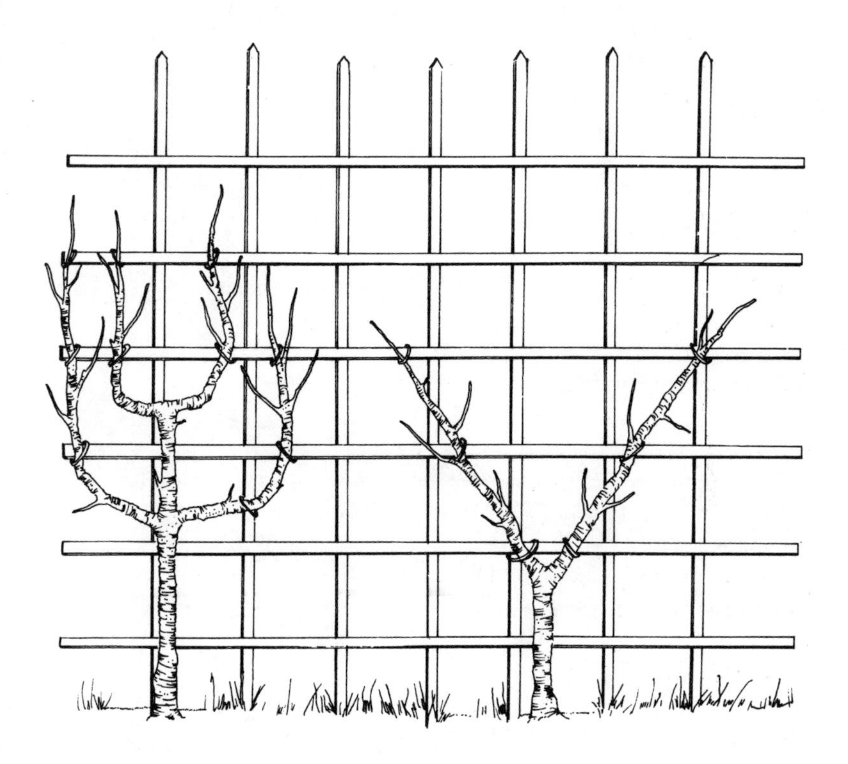

Here are two typical examples of fruit trees starting to be trained in the espalier system. At left is the candelabra form, at right the classic V form.

Some patterns fit better in certain areas than others. The cordon design, for example, is effective against wide expanses of wall. It looks good beneath high windows, especially on tall walls because it provides a horizontal effect that helps bring the too-tall wall back into more pleasing perspective.

You can use any design that catches your imagination. The simpler the design, of course, the easier it will be to maintain. Study the background wall or area. Visualize which design will look best. Consider the background color of other foliage in adjacent properties and

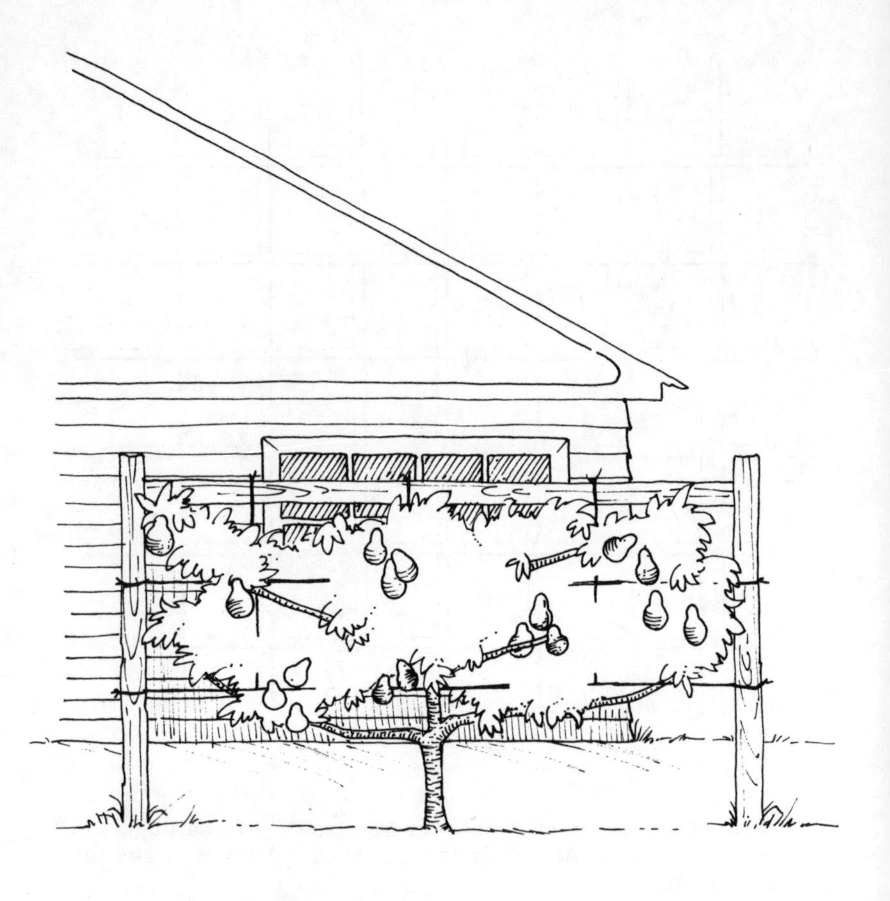

Trees pruned in the espalier fashion, like this pear tree, may yield slightly less fruit than the conventional tree, but it can also serve as a screen.

the need to paint and maintain walls behind your proposed espalier plants.

Try a sketch on paper first. Draw the background wall. Then trace on different designs to see which appeals most. Keep in mind windows that must be reached and power lines that may interfere as plants mature.

Although espalier is generally meant to grow in one plane, it is possible to train trees around a corner. By planting four trees in a square with a lattice roof, you can also create a coolly shaded sitting area. When plants mature, their top branches become your roof. Sitting beneath such a living garden arbor, we watched a friend reach up to pick an apple. His outdoor living room certainly had its crisp rewards.

PLANTING AND SUPPORTING ESPALIERS

You should plant your selected fruit trees as you would those intended for other parts of your home grounds. Dig the hole deep enough. Improve the soil by mixing humus, well-rotted manure, peat, or compost with the soil in whatever proportions are needed.

Before you plant, plan and erect your espalier supports. Reinforcing steel rods used in construction or sturdy posts to which you anchor training cables or heavy wires should be positioned well. Some can be removed in the future once your trees take proper and desired shape. Don't set them in concrete. They'll be needed for only a few years.

But in training grapes to espalier form, install permanent supports. The same is true for blackberries and raspberries. These too can be well tamed to espalier habit, but cannot support themselves without drooping. They'll need permanent supports with wires or other props.

The simplest design to create is a fan or hand. You can secure rods, posts, or poles at least a foot, perhaps 18 inches, deep in the soil. Be sure to keep them away from the wall since you do need room to prune behind as well as among the branches of your chosen plants.

If you use wires, take up the slack so they don't sag and whip around in the wind. Loose wire can whip into tender bark, cutting and scarring it. The result is damaged spots that disease and insects can penetrate.

Once the basic support is in place, plant your espalier-to-be tree, shrub, or vine. When it is firmly planted properly, next step is tying. Attach limbs to the support with short lengths of soft twine. You can also use weather-resistant wire that's coated with paper or plastic. Uncoated wire will too easily cut into bark and girdle stems. However, wire run through plastic fish-tank hose or similar material serves well.

Branches can be bent a bit to achieve the desired effect. But heed this caution: *don't* try to accomplish complete shaping all at once! Branches may be flexible, but they will break if bent too far. Give them time to turn in your proposed direction naturally. You can take up the slack in a month or so, leading them in your preferred pattern step by growing step.

Once you have attached the desired branches, step back and examine your progress. Plants don't grow perfectly in balance, so don't be disappointed. A bit of careful pruning of a growing tip here, an unwanted minor branch there, will let you mold your living sculpture to your own desires.

If your espalier will be against a wood or even a brick wall, you can possibly do without a free-standing support. Eye hooks or J-hooks can be screwed to the wall at appropriate points. Limbs and branches can be tied to them, not too tightly nor too loosely. Ice storms do occur in some areas and can break branches on your espaliers.

When you must maintain the wall, painting or touching up cement, the connecting cord or coated wire can be loosened. Then gently pull the espalier over and tie until the paint is dry.

On brick or concrete walls, special nails can be driven into the mortar. A soft lead tongue is attached to the head of these nails. This can be bent over to secure the branch.

You can also establish your espalier on a frame of heavy wire such as modern plastic-coated clothesline. If the white color offends you, paint it green to blend with and virtually disappear into the foliage.

PRUNING YOUR ESPALIER

Once you have your frame in place, your espalier planted well and secured to it, think about the shapes to come. Keep pruning shears handy, but don't be too hasty. Once you cut away a branch or branchlet, you encourage others to form. Also, once that one is gone, another seldom will sprout there.

Watch for branchlets that grow parallel to the ground and for those that grow toward or away from the wall. They'll spoil the espalier look quickly if you don't remove them.

The best time to prune is when flower buds form. Then you can see where future blooms and fruits or berries will be and avoid cutting them away. Each side or "lateral" shoot that forms along the main stem should be removed at the main stem if it is not part of the desired pattern.

When new growth appears in unwanted places, pinch off or clip away the tip to prevent stray branchlets from growing. With a little practice, you'll soon get the feel of plantscaping to the living sculpture forms you want.

Root pruning is often overlooked. Many espalier trees and shrubs grow vigorously. They can outgrow even your dedicated pruning effort. Root pruning checks growth and also encourages flowering and fruiting.

To do this, simply dig a shallow trench around the plant to cut away some large thick roots. Do this judiciously, since you only want to keep the plant in check, not destroy too much of its food- and water-gathering root system. Make your trench about 2 feet away from the trunk. Sometimes you can root prune by spading around low-growing espalier shrubs.

Fertilize your espalier fruit trees and berry bushes somewhat less than you would if they were left to grow to their natural full size. Tips for feeding are included in the appropriate chapters of this book. Deep-root feeding with fertilizer spikes placed in hand-driven holes around the trunks, or root feeding devices like the Ross Root Feeding tool used with a garden hose, can give espalier plants a boost in season.

This Old World plantscape technique lets you create showy and unique accents for your home grounds. You'll find that dramatic new shapes lend a distinctive and personal look to your outdoor living areas. The nicest part about espalier trees and shrubs is that once you have established them, just minimal pruning each year keeps them in bounds. For little care, they'll give you back lots of pleasure that is distinctively your own creation.

17. Know Your Insect Enemies and How to Beat Them

Whether we like it or not, insect enemies and plant diseases can attack even the best-tended fruit trees and bushes. Although it's true that healthy, vigorous plants have some ability to withstand and fight off disease and insect problems, at times a combination of conditions favors the pests and not your plants.

Before that time comes, it helps to know some of the arch villains that may visit your home fruit planting. They're not particularly nice to know. But if you can spot these problems and pests before they launch full-scale attacks, they are easier to defeat.

Prevention of problems before they occur has its obvious advantages. Chemical sprays, whether you approve of them or not, can prevent most insect and disease problems. On balance, considering the value of the trees and plants and in keeping with sound balance in nature, there are ways to select the better, safer-to-use materials that stop pests yet have few if any adverse effects on people or the environment.

A basic pest-prevention plan—including suggested materials to use and when to use them—is included in the next chapter. It is designed as a general guide. Local areas have special problems. County agents and your supplier of garden products can provide local details.

Here's the rogue's gallery. Being able to identify them will help you to banish them before they eat your fruitful harvests.

Aphids (plant lice) attack foliage of many types of fruit. These small, soft-bodied insects suck juice from leaves, causing them to crinkle or curl. Aphids are usually green, brown, or black.

Cankerworms and other loopers feed on the leaves and young fruits of many trees. When disturbed, they drop and hang suspended on strands of silk. They usually disappear by late spring or early summer.

Codling moths are responsible for wormy apples and pears. The larvae eat through the flesh of apples, leaving tunnels filled with brown debris.

Currant fruitflies lay eggs under the skin of currant and gooseberry fruits. Larvae feed on fruit seeds and pulp, which turns red and drops to the ground.

Cyclamen mites are tiny pests. They live in the crown of strawberry plants and attack leaves and flower buds. Leaves become wrinkled and brown at the tips. Mites may prevent fruit formation or cause misshapen fruit.

Fall webworms develop in large web nests on the branches of fruit trees during late summer or fall. They eat leaves in and around the nest. Prune off branches and destroy the nests as soon as they appear.

Grasshoppers often strip trees of foliage and tender bark. When hoppers become numerous, control them in ground cover and fence rows to prevent hoppers from migrating to fruit trees. If hoppers hit fruit trees, treat directly.

Leafhoppers are green, gray, or tan, about ⅛ inch long. They feed on the undersides of leaves of many fruits. Young hoppers suck juice from leaves. They may transmit diseases from one tree to another.

Lesser peach tree borers attack stone-fruit tree trunks and scaffold branches, particularly where pruning wounds and other injuries make penetration easy. Look for a gummy exudate.

Peach tree borers work at the base of stone-fruit trees. Sawdust castings and gummy masses at the base of the trunk indicate their presence.

Plum curculios are snout beetles. They deposit eggs in young tree fruits. Egg punctures are crescent-shaped rather than round as with the plum gouger. Damaged fruit generally turns red prematurely and falls from the tree.

Plum gougers are another type of snout beetle. They deposit eggs in stone fruits while the fruits are small. Larvae burrow into the pit and emerge as adults, leaving a matchstick-size hole in the fruit. Try this: spread a canvas in early morning under the infested tree. Strike the trunk with a padded mallet, or shake vigorously. Destroy all beetles that drop onto the canvas. Repeat at four-day intervals until no more beetles are found.

Raspberry fruitworms. Adults are small light-brown beetles about ⅛ inch long. They feed on young buds, leaves, and blossoms. Yellowish-white larvae live in or on ripening fruit.

Raspberry sawflies are black, thick-bodied insects about ¼ inch long. They lay eggs on raspberry plants when raspberry leaves begin to unfold. Small, spiny, green larvae feed on the leaves.

Scale insects are small sucking insects about 1⁄10 inch long. Except for a brief period when the young are hatched, scale insects remain attached to a branch or twig. *Oystershell scale* is brownish and has a shape similar to that of an oyster shell. *Scurfy scale* is white and pear-shaped.

Spider mites are tiny insectlike pests that suck juices from leaves, which eventually become bronzed and dried and fall. Spider mites frequently are troublesome during dry years. Their presence may be detected by fine webs on the undersides of leaves. Hosing trees with a garden hose often aids in control.

Strawberry weevils are small reddish-brown snout beetles up to ⅛ inch long. They feed on stems of fruit buds, which wilt or drop off.

Tarnished plant bugs are brownish bugs about ¼ inch long. They feed on strawberries and other plants, damaging foliage and fruit. Both adults and nymphs injure plants.

White grubs are the large, thick-bodied larvae of May beetles, also called June beetles in northern states. Larvae feed on roots of plants, including strawberries. Avoid planting strawberries in newly turned sod.

RECOGNIZE DAMAGING DISEASE

Diseases can sometimes cause as much damage as insects do. In fact, when some diseases are severe they can defoliate trees and cause the dropping of most of the fruit. Here's what the most common are like so you can spot them early and apply necessary fungicides to control the problem.

Anthracnose is a fungus disease of raspberries. It affects purple and black varieties most. It is characterized on canes and leaves by round or oval spots with a brown, red, or purple border and gray center. Spots are up to ½ inch in diameter. Clean cultivation, which promotes air circulation, helps control this disease.

Apple scab is a fungus disease that infects the leaves, twigs, and fruit of apple trees. This disease appears on leaves as dark-green velvety spots. On fruit, the disease first appears as a slightly raised brown or black round spot that later breaks open to form a scab. Infected fruits frequently crack or become malformed. Fungi overwinter in fallen leaves and reinfect trees during rainy weather in spring.

Keep trees well pruned so that sunlight and wind will dry the leaves quickly after rain. Also clean up and, where permitted, burn leaves.

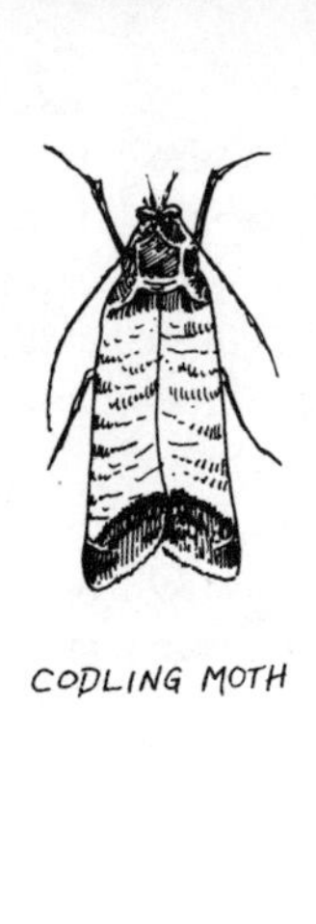

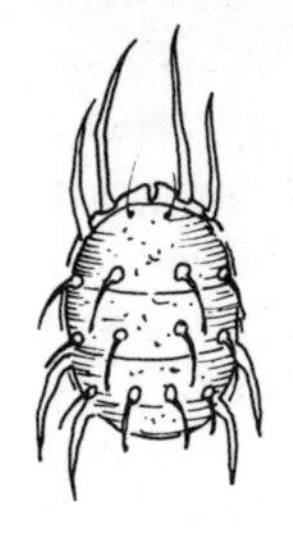

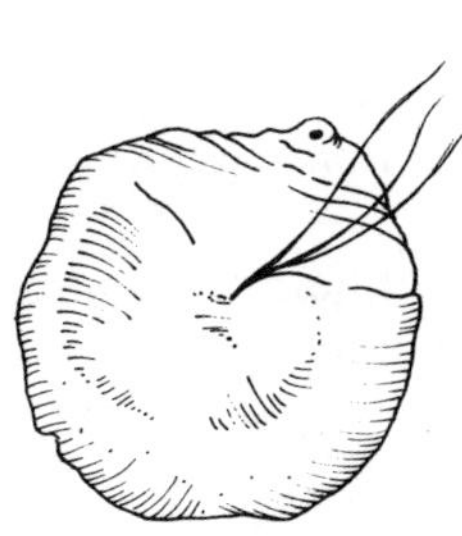

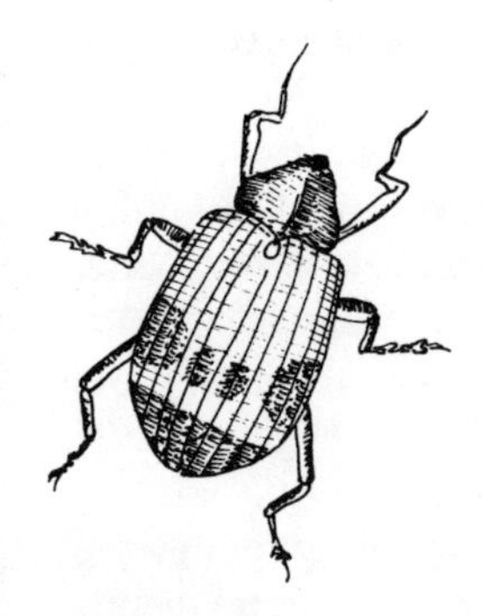

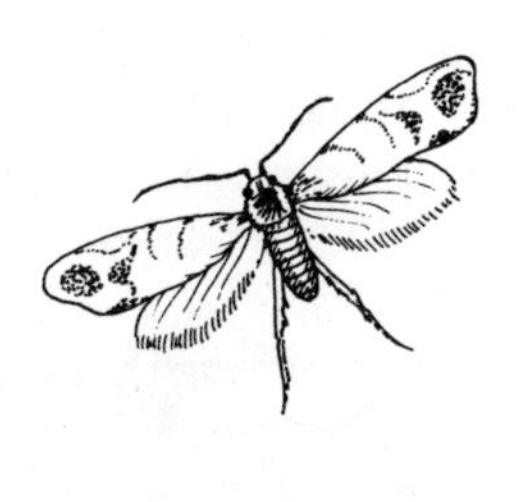

These nine are among the most common enemies you may detect in your fruit trees or berry bushes. Identification is first step toward control.

Brown rot of stone fruits is primarily a fruit disease but also infects leaves, flowers, and twigs, giving them a frosted look. If ripening fruit is infected, brown spots will form and enlarge until the entire fruit becomes soft, watery, and discolored. Fruits later become covered with brown tufts, dry up, and hang like mummies on the tree. These should be removed and burned because the disease overwinters in them.

Cedar rust, often called apple rust or cedar apple rust, is a fungus disease that originates on cedar trees and infects apple trees. On cedar trees the disease appears as a brown corky gall. During spring rains, galls swell to become an orange-colored, jellylike mass, which contains spores that infect fruit trees.

On apple trees this disease appears first as yellow spots on the upper leaf surface. These spots increase in size and become orange with small black specks in centers. Leaf tissue beneath spots swell into blisters with tubular projections. Severe infections may cause leaf drop and deformed fruit. If possible, remove red cedar trees from within a ¼ mile radius of the apple trees.

Fire blight is a bacterial disease that may severely injure susceptible varieties of apples and pears. Pollinating insects transmit the disease to blossoms, and sucking insects transmit it to growing shoots. The disease normally starts in growing tips and progresses downward, killing all tissue it invades. Infected leaves become brown or black, dry up, and remain attached to branches. Infected twigs become dark. Bark is sunken. A milky or brownish ooze may form on the infection. Cankers form where the disease enters a large branch or trunk. The bacteria overwinter in these cankers and are a source of infection the following year.

As soon as diseased shoots are noticed, remove them by breaking or cutting them off several inches below the infection. If a cutting tool is used, it must be disinfected in strong household bleach—after every cut—so that bacteria will not be spread to healthy tissue.

Remove suckers and watersprouts that grow from the trunk and large branches. These sprouts are highly susceptible to blight. Disease readily passes through them into older wood.

Leaf spot is a common disease in currants. Infected leaves develop numerous small round spots with gray centers. This disease frequently causes leaves to fall early. Rake and, if possible, burn them.

Plum pockets is a fungus disease that infects the fruits, shoots, and leaves of plums. Infected fruits and shoots puff up into large hollow masses and eventually drop.

Powdery mildew infects many plants and is occasionally serious on currants and cherries. It causes a white moldy growth that distorts leaves and stem tips.

Spur blight in raspberries is a fungus disease characterized by purplish oval spots around the buds. Infected areas are weakened, and fruit production is reduced.

Sunscald is a winter injury rather than a disease. In late winter or early spring the bright afternoon sun warms the southwest side of trees and activates cells under the bark. These cells are then killed by later freezing. Bark sloughs off, leaving a canker that will weaken and eventually kill the tree. Sunscald usually injures the trunk and large limbs.

Sunscald can be prevented by several methods of protection. You can lean boards over the southwest side of the tree to shade it from the sun, wrap the tree trunk with burlap, tar paper, or aluminum foil, or whitewash the trunk and large limbs to reflect the sun. Natural color tree wrapping is available at garden centers; it is unobtrusive and works well.

18. Pest-Control Guidelines

When to spray or dust is just as important as what to apply in order to control harmful insects and plant diseases on your fruit trees and berry plantings. Proper timing is important if you hope to solve the problems and provide adequate protection for the plants between treatments.

Some areas have more insect problems than others, and sometimes different types of pests. Seasons are different too. One year you may have lots of rain, which requires more attention to cover sprays to prevent mildews, spots, and rusts on trees and fruit bushes or vines. If you have just sprayed and you get heavy rains, you may need to spray again after the rain to reapply material washed off the foliage, leaving it unprotected.

Every county agent has a suggested spray schedule that has been tested in his area and state. Your best bet is to check with your county agricultural agent for the timing, recommended materials, and rates proved best for the chemicals most effective and on your state's approved list.

A general guide, however, is in order. This guide will provide the general procedure for when and how often you should plan to protect your fruit crops from damage by insects and plant diseases. You may prefer different materials. Consult your local pesticide supplier for details about the materials that have proved most effective in the specific climate, soil, and other conditions of your area.

Here's the general guide which we have found most useful in plotting our campaign of pest control and prevention.

APPLES AND PEARS

Dormant Spray

For scale, aphids, and red mites, use a Dormant (or Miscible) Spray Oil. Apply in spring before green begins to show in buds.

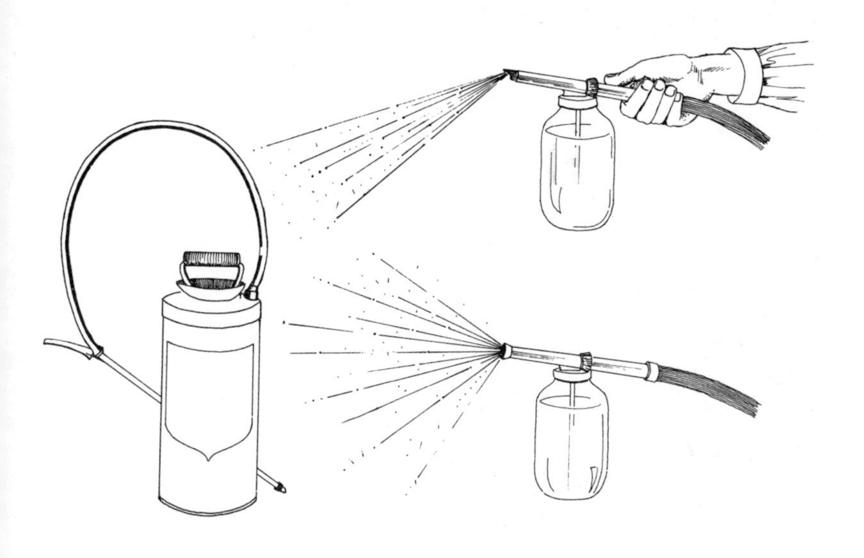

Three of the ways you may choose to apply sprays to your fruit trees or berry bushes: left, tank pump sprayer; top right, directed spray; bottom right, adjustable fine-mist spray.

Delayed Dormant Spray

For apple scab and mildew, use Microfine Wettable Sulfur. Apply when unfolding leaves are ¼ to ½ inch long.

Pre-Bloom Spray for Apples

(May be omitted on pears.)

For apple scab and mildew, use Microfine Wettable Sulfur. Apply when most of the blossom buds are beginning to show color and the individual buds have separated into clusters.

Bloom Spray

For scab, mildew, and cedar rust, use Microfine Wettable Sulfur plus Zineb or Mancozeb. Apply when about one-quarter of the blossoms are open.

Petal Fall Spray

For scab, cedar rust, codling moth, curculio, and leafrollers, use Captan-1 plus Zineb or Mancozeb plus Methoxychlor-1. Apply after most of the petals have fallen.

First Cover Spray

For scab, cedar rust, codling moth, curculio, and leafrollers, use the same materials as for petal fall spray, below. Apply about seven to ten days after the petal fall spray.

Second Cover Spray

For scab, cedar rust, codling moth, curculio, and leafrollers, use Captan-1 plus Zineb or Mancozeb plus Diazinon. Apply about seven to ten days after the first cover spray.

Third, Fourth, and Fifth Cover Sprays

For scab, codling moth, leafrollers, and mites, use Captan-2 plus Diazinon. (Do not use Diazinon within fourteen days of harvest) Apply at intervals of seven to ten days.

Summer Cover Sprays

For scab, summer diseases, codling moth, leafrollers, and mites, use Captan-1 plus Zineb or Mancozeb plus Carbaryl plus Kelthane plus Malathion-1. The number of cover sprays needed will depend on the maturity date of the variety. Summer cover sprays may be applied at intervals of ten to fourteen days. DO NOT apply this combination on apples or pears within seven days of harvest. Read label details on closeness of applications to harvest.

CHERRIES

Dormant Spray

For overwintering pests and scales, use Dormant (or Miscible) Spray Oil. Apply in the spring before any green is showing in the buds.

Petal Fall Spray

For mildew, leaf spot, brown rot, and curculio, use Folpet plus Captan-1 plus Methoxychlor-2. Spray when most of the petals have fallen.

First Cover Spray

For mildew, leaf spot, brown rot, and curculio, use the same materials suggested for petal fall spray. Apply ten days after petal fall spray.

Second Cover Spray

For mildew, leaf spot, brown rot, and curculio, use Folpet plus Captan-1. Apply about ten to twelve days after the first cover spray.

After-Harvest Spray

For mildew and leaf spot, use Folpet plus Captan-1. Apply as soon as all the fruit has been picked. During wet summers, apply a second after-harvest spray about three to four weeks later.

PEACHES, PLUMS, AND APRICOTS

(Substitute Captan for Sulphur on Apricots)

Dormant Spray

For scale insects and peach leaf curl, use a dormant spray every year. Use Dormant or Miscible Spray Oil plus Ferbam-2. Apply early in the spring, shortly before any growth begins. Mix the spray oil and Ferbam together before adding water. Shake or stir vigorously, then add a small quantity of water and stir again, after which the combination should mix with the remainder of the water.

Pre-Bloom Spray

For brown rot and catfacing insects, use Microfine Wettable Sulphur plus Carbaryl (Sevin). Apply when blossom buds show pink.

Bloom Spray

For brown rot, use Microfine Wettable Sulphur. Apply when about a quarter of the blossoms are open.

Petal Fall Spray

For brown rot, scab, and curculio, use Microfine Wettable Sulphur plus Methoxychlor-2. Apply when most of the petals have fallen.

Shuck Spray

For brown rot, scab, curculio, and Oriental fruit moth, use the same material as for petal fall spray. Apply about ten days after petal fall spray.

First, Second, and Third Cover Sprays

For brown rot, scab, curculio, and Oriental fruit moth, use Microfine Wettable Sulphur plus Carbaryl. Apply about ten days after shuck spray and at ten-day intervals.

Summer Cover Sprays

For brown rot, Oriental fruit moth, and mites, use Captan-2 plus Carbaryl plus Kelthane. The number of cover sprays needed will depend on the maturity date of the variety. Apply summer sprays at ten-to-fourteen-day intervals. DO NOT use this combination within two weeks of harvest.

Pre-Harvest Sprays

For brown rot and mites, use Captan-2. If mites are present, add Malathion-1, but not within seven days of harvest. Apply one spray about two weeks before harvest and a second spray about one week before harvest.

Special Summer Sprays

For peach tree borer, use Methoxychlor-2. Make three applications: (1) the first week of June; (2) the first week of July; (3) the first week of August. DO NOT use within twenty-one days of harvest. Apply as a spray or with a brush to the trunks of the trees from the ground line up to about one foot on the trunk. Permit the material to run down the trunk and soak into the ground.

GRAPES

First Spray

For black rot and grape flea beetle, use Ferbam-2 plus Carbaryl. Apply when most of the new shoots are about one inch long.

Second Spray

For black rot, mildew, and grape flea beetle, use Ferbam-1 plus Folpet plus Carbaryl. Apply about ten days after the first spray.

Third Spray

For black rot and mildew, use Ferbam-1 plus Folpet. Apply ten days after the second spray.

Fourth Spray

For black rot and mildew, use the same materials as for the third spray. Apply as blooming begins.

Fifth Spray

For black rot, mildew, grape berry moth, and mealybugs, use Folpet plus Ferbam-1 plus Malathion-1 plus Carbaryl. Apply as soon as blooming is complete, or about ten days after the fourth spray.

Sixth Spray

For black rot, mildew, berry moth, mealybugs, and scale, use the same materials as for the fifth spray. Apply about two weeks after the fifth spray.

Seventh Spray

For black rot, mildew, and berry moth, use the same materials recommended for the fifth spray. DO NOT use Malathion-1 within three days or Ferbam-1 within seven days of harvest. Apply about two weeks after the sixth spray.

GOOSEBERRIES AND CURRANTS

Dormant Sprays on Currants

For scale insects, use Dormant or Miscible Spray Oil. Apply in spring before any new growth starts.

Delayed Dormant Spray

For leaf spots, use Ferbam-2. Apply just as the leaves appear and start unfolding in the spring.

First Cover Spray

For leaf spots, aphids, and currant worm, use Ferbam-2 plus Malathion. Apply about ten to twelve days after the delayed dormant spray.

Second Cover Spray

For leaf spots, use Ferbam-2. Apply two weeks after the first cover spray, but DO NOT use Ferbam-2 within fourteen days of harvest.

Special Spray

Watch the interior of the bushes for the feeding of currant worm. Use Malathion-1 after the fruit has developed and before picking. DO NOT use Malathion-1 within three days of harvest.

After-Harvest Spray

For leaf spots, use Ferbam-2. Apply after the completion of fruit harvest.

RASPBERRIES, DEWBERRIES BOYSENBERRIES, AND BLACKBERRIES

Delayed Dormant Spray

For anthracnose on old canes, use Liquid Lime Sulfur. Apply when the buds break in the spring—just as the green leaf tips appear.

Second Spray

For anthracnose on newly developing canes and old canes too, use Ferbam-2 or Captan-2. Apply ten days after the delayed dormant spray.

Third Spray

For anthracnose on newly developing canes and old canes, use the same materials as for the second spray. Apply ten days after the second spray. DO NOT use Ferbam-2 within forty days of harvest.

Additional Sprays

For anthracnose on newly developing canes and old canes, use Captan-2. Continue spraying at ten-day intervals until the young fruits appear.

Special Spray for Spider Mites

For spider mites and chiggers, use Malathion-1. Apply when mites appear, but DO NOT use Malathion-1 within one day of harvest.

STRAWBERRIES

Pre-Bloom Sprays

For foliage disease, fruit rot complex, catfacing insects, weevils, rootworms, and crown borers, use Captan-3 plus Malathion-2. Apply the first spray when plants resume growth early in the spring—just as soon as the mulch is removed. Spray over all the plants and all the ground area on the rows and between the rows. Repeat applications of the same materials at seven-to-ten day intervals *until blooming begins*. DO NOT SPRAY DURING BLOOM.

After-Harvest Sprays

Leafrollers may become a problem after harvest, use Malathion-1. Begin application when the problem is observed, and continue at two-week intervals until it is brought under control.

Spider mites become a problem during the hot, dry summer too. Use Kelthane. Begin spraying when mites are first noticed, and make the second application in ten days.

19. Pruning Pointers

Begin training a young fruit tree the day you plant it so that it will develop strong branches and good form for heavy fruit production.

Most fruit trees, if they're not pruned, develop an umbrella shape. This shape is bad because outer leaves shade inner parts of the tree and you get less fruit. Umbrella trees also have many weak limb crotches. They'll tend to break apart in storms or when heavy with fruit.

Proper training while the tree is young, plus regular pruning later, produces a desirable strong, spreading, open-centered crown that lets light into the tree. It is important to let the sun shine in.

Fruit trees, as they come from reliable nurseries, usually have many more branches than they need. About 75 percent of the plant's roots are destroyed when it's dug from the nursery. This reduced root system will adequately support only a correspondingly reduced number of branches. That means, you must remove a goodly amount of the branches to compensate for root loss.

To begin training apples and pears, remove all side branches for 2 to 3 feet up the trunk. The first branch at this level that forms an angle of 45 to 90 degrees with the trunk should be left on. For greatest crotch strength, all the major side (scaffold) branches should make angles of about 45 to 60 degrees with the trunk. Limbs that make smaller angles develop weak crotches because the limbs and trunk become pressed together as they grow.

After you choose the lowest scaffold branch, select four to six other wide-crotch-angle branches spaced 8 to 12 inches apart in a spiral up the trunk. Keep these, and remove all the others. Don't select a branch that is directly above another. It will shade the lower branch. Cut branches off flush with the tree trunk. Don't leave stubs. They heal slowly and invite disease.

Then cut each scaffold branch back by one-third its length. Finally, cut back the leader (top extension of the trunk) so it extends slightly above the side branches.

Smaller trees such as dwarf apples, plums, and cherries are trained in about the same way. However, the four to six scaffold branches can be closer together and closer to the ground.

Even if you planted fruit trees a few years ago, there's still time to train them. Trees that have already begun bearing can also be pruned in early spring to keep them strong and healthy.

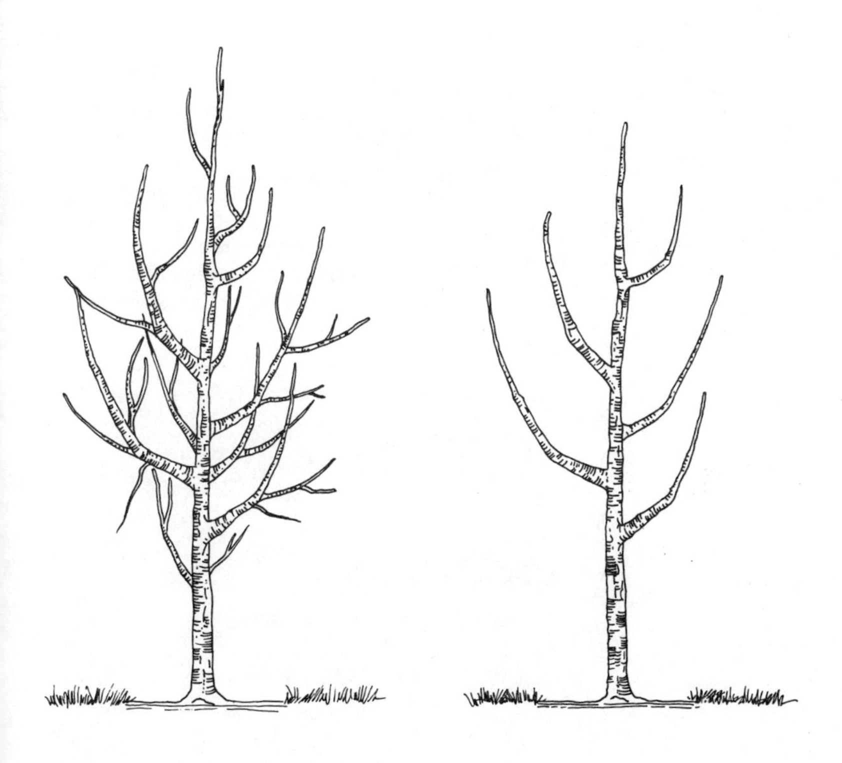

An apple or pear tree that is three to five years old may become scraggly (left). Proper pruning can get it back under control (right).

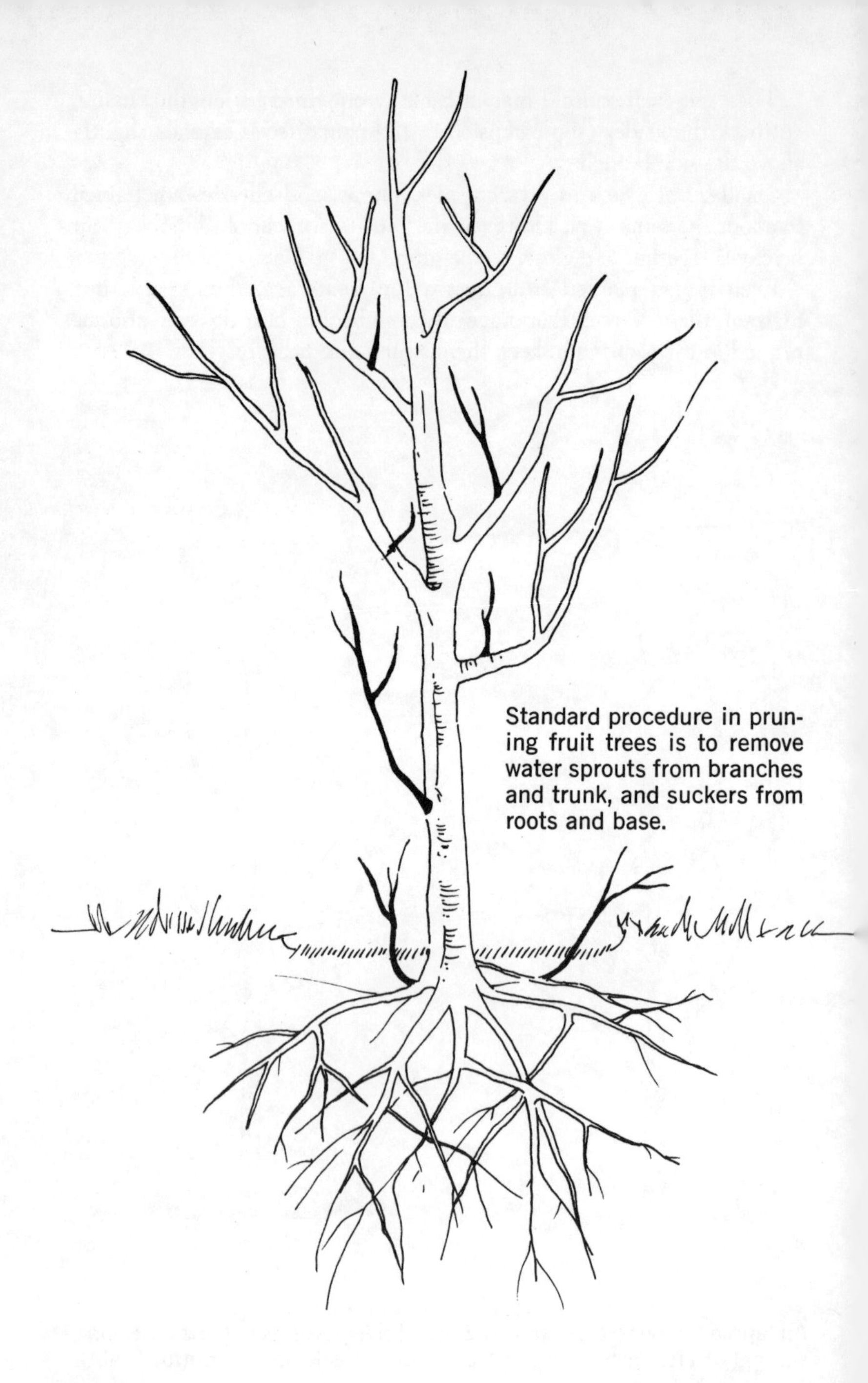

Standard procedure in pruning fruit trees is to remove water sprouts from branches and trunk, and suckers from roots and base.

If double leaders have developed at the top of the tree, remove one of them. Prune out any water sprouts. These are those vigorous, upward-growing shoots rising from the trunk or scaffold limbs. Also remove suckers. Those are the shoots rising from the base of the trunk. Neither performs any useful service.

Remove one of any two branches that conflict or rub together. Also prune away any dead branches, all branches with weak crotch angles, and any branches that grow toward and shade the tree's center. After a number of years, you may want to cut out the leader to limit the tree's height. Those are the basics.

Pruning can be one of the most important aids to peak fruit production, especially from tree fruits. It pays to use well-sharpened tools. Here are the most important points for proper pruning.

Smooth, clean cuts can be made only with sharp tools. Clean cuts

From left to right you see typical recommended pruning for an apple or pear tree in its second year, its third year, its fourth year, and fifth or sixth year.

heal more quickly than ragged ones. Insects and diseases have less chance to penetrate properly made pruning cuts.

Make cuts flush. Decay often starts in ragged cuts and may then enter the main trunk to weaken the entire tree structure.

Place the cutting blade of your shears below or at the side of the crotch. Never place it in the crotch of the tree. Do and don't illustrations show the right and wrong way.

Cut evenly. Don't wiggle the shears through a cut or you may spring the shears or leave a ragged wound. If you prune regularly, you will find it easier to remove small branches and suckers rather than having to saw away larger limbs.

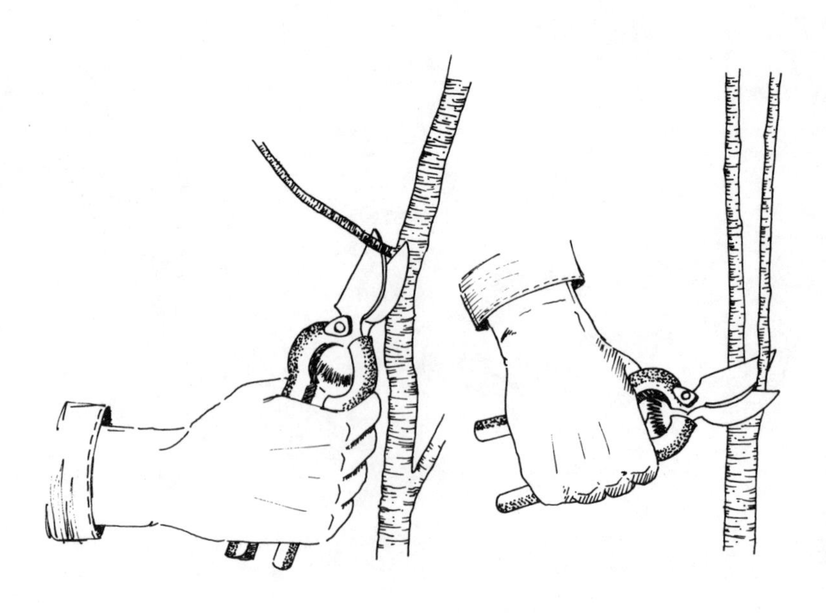

Even with something as seemingly simple as using pruning shears, there's a right way (left, blade at side or bottom) and a wrong way (blade on top, in crotch).

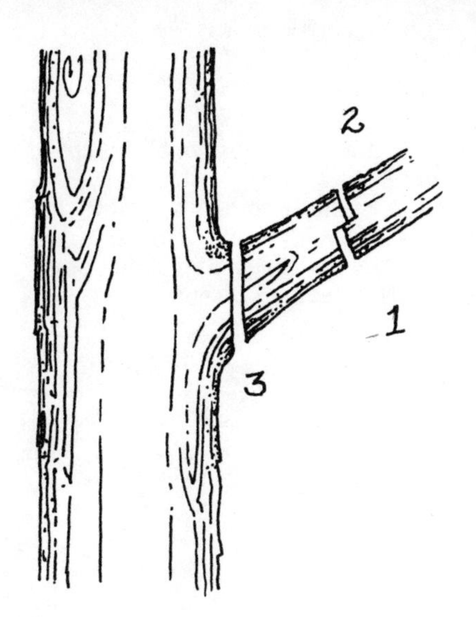

The right way to remove a heavy limb involves three steps: 1) saw upward about 6 inches from trunk; 2) saw down about 8 inches from trunk; 3) saw stub off flush to trunk.

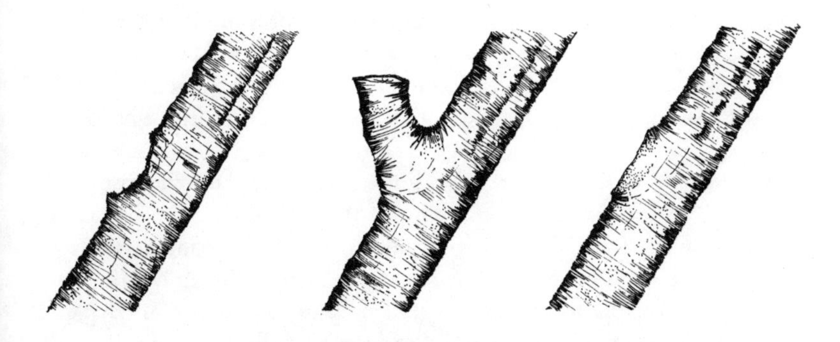

Branches too big for pruning shears must be removed with a saw. Here, from left to right, are cuts too close to a main limb, too far away, and just right (flush).

If you must remove a large branch or limb, here's how to avoid tearing the trunk bark:

Make a 2-inch-deep cut in the bottom of the limb, about 6 inches

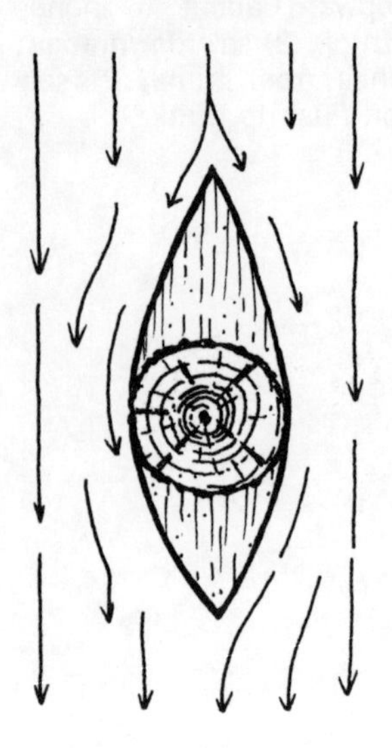

You can give a professional touch to the trunk area from which you've removed a stout limb by trimming the bark in elliptical shape. This way you thwart standing water and deter penetration by insects or disease.

from the base. Next cut down into the limb about 8 to 10 inches from the trunk. When the limb falls, the undercut prevents tearing of the bark down along the trunk itself. Finally, use a sharp saw and remove the stub as flush to the tree as possible without cutting into the trunk itself.

When you must remove large limbs that have become diseased, damaged, or broken by storms, remove them cleanly. Any branches or limbs more than 2 inches in diameter should be painted with a standard tree-wound paint. You can get such paint under several brand names in hardware stores, garden centers, and garden departments of some chain stores. Mail-order firms also sell it.

With large cuts, it helps to finish your pruning as a professional tree surgeon would, leaving an elliptical shape. This lets water drain away, rather than catch in a rough or ragged spot where fungus or other problems can start. Use a sharp knife or chisel and shape the wound from the removed limb as shown in the illustration. It will heal more quickly and more surely than a sloppy wound and will prevent problems. Coat or paint these larger wounds with tree-wound paint.

Handy Quick Reference and Information Sources

BASIC THUMBNAIL GUIDE TO FRUITS AND BERRIES

Sometimes it helps to have an overall view as we plan our long-range garden plantscape. Here's a handy thumbnail guide to distances between rows and individual plants in berry patches and home mini-orchards. If you interplant with other landscape trees and shrubs it works about as well. Just be sure to provide enough growing room for trees and bushes and room to roam for vines and plants that ramble in a row or brambles.

Fruit	*Distance in Feet Between*		*Usual Bearing Age*	*Approximate Yield and Period*
	Rows	*Plants*		
Apple	20–25	35	6–8 years	3–6 bushels, Aug.–Oct.
Dwarf Apple	15–20	8–10	4–5 years	2 bushels, Aug.–Oct.
Dwarf Pear	15–20	10–20	3–4 years	1 bushel, Aug.–Sept.
Peach	15–20	20	3–4 years	2 bushels, July–Sept.
Cherry, sweet	20–25	25	6–7 years	1 bushel, July
Cherry, sour	15–20	20	4–5 years	1 bushel, July
Plum	15–20	20	4–5 years	1 bushel, Aug.–Sept.
Currant	8	3	2–3 years	3 quarts, June–July
Blueberry	6	4	3–4 years	2 quarts, July–Aug.
Grape	10	10	3–4 years	8 pounds, Sept.–Oct.
Raspberry	8	3	2–3 years	1 quart, July–Sept.
Blackberry	8	3	2–3 years	1 quart, July
Strawberry	3	2	1 year	1 pint, June–Sept.

GUIDE TO THINNING

Thinning is necessary to ensure that trees put their growing strength into a selected amount of fruit. That way, these fruits become plumper, juicier, and tastier. Excess fruits can be set at certain times. Too many left on trees to mature will result in smaller, less-desirable fruits come harvest time. Here's a handy guide for hand-thinning to let your trees produce the best possible fruit to perfection. Remove fruits to the distance suggested along each branch.

Apple	6 to 8 inches
Apricot, Peach, nectarine	2 to 3 inches
Cherry	1 inch
Pear	4 to 5 inches
Plum	1 inch for small
	2 to 3 inches for large

WHAT WILL THEY YIELD?

It's nice to know before you grow what your bush fruits may yield for you. Here's another handy chart which I've found is quite accurate. It has helped as I plan new plantings of various bush fruits, grapes, and strawberry plants.

Fruit	*Minimum Distance Between Rows*	*Minimum Distance Between Plants*	*Average Annual Yield per Plant*	*Bearing Age*	*Life Expectancy*
	feet	feet	quarts	years	years
Blueberry	6	4	4	3	20–30
Blackberry (erect)	8	3	1½	1	5–12
Blackberry (trailing)	8	6	1½	1	5–12
Raspberry (red)	8	3	1½	1	5–12
Raspberry (black)	8	4	1½	1	5–12
Raspberry (purple)	8	3	1½	1	5–12
Grape (American) (Fr. American)	10	8	15 lbs.	3	20–30
Grape (muscadine)	10	10	25 lbs.	3	20–30
Strawberry (regular)	3	1	½	1	3
Strawberry (everbearing)	3	1	½	1/3	2
Currant	8	4	5	3	10–20
Gooseberry	8	4	5	3	10–20

A WORD ON FREEZING, CANNING, JAMS, JELLIES, AND PRESERVES

This book is not intended as a guide to freezing, canning, or making jams, jellies, or preserves. However, since many gardeners like to know what quantities of different types of fruit convert to quantities for their home freezer or for estimating yields for other uses, here's a list I've used for years. It varies, naturally, with size of fruit, stage of ripeness, and variety. However, it gives you a guide to planning for ways to preserve the fruits of your gardening fun.

Fruit	*Amount*	*Yield*
Apples	1 bushel	32–40 pints
	1¼ lbs.	1 pint
Berries*	1 crate (24 quarts)	32–36 pints
	1½ pints	1 pint
Cantaloupes	1 dozen	22 pints
	1–1¼ lbs.	1 pint
Cherries	1 bushel	36–44 pints
	1¼ lbs.	1 pint
Cranberries	1 25-lb. box	50 pints
	½ pound	1 pint
Currants	2 quarts	4 pints
	¾ lb.	1 pint
Peaches	1 bushel	32–48 pints
	1–1½ lbs.	1 pint
Pears	1 bushel	40–50 pints
	1–1¼ lbs.	1 pint
Plums	1 bushel	38–50 pints
	1–1½ lbs.	1 pint
Raspberries	1 crate (24 pints)	24 pints
	1 pint	1 pint
Rhubarb	15 lbs.	15–20 pints
	1 lb.	1 pint
Strawberries	1 crate (24 qts.)	38 pints
	2/3 quart	1 pint

*Berries indicated in this chart include blackberries, blueberries, dewberries, elderberries, gooseberries, and huckleberries.

REPUTABLE SUPPLIERS OF FRUIT TREES, SHRUBS, AND PLANTS

Here's a list of leading nurseries and mail-order firms that have been supplying fruit trees, shrubs, and plants to American gardeners for many years. Their reputations are well established. Many of these firms have beautifully illustrated, colorful catalogs for free or at nominal cost.

Bountiful Ridge Nurseries
Princess Anne, Maryland 21853

W. Atlee Burpee Co.
300 Park Avenue
Warminster, Pa. 18974

Burgess Seed and Plant Company
Galesburg, Michigan 49053

Cumberland Valley Nurseries, Inc.
McMinnville, Tennessee 37110

Emlong Nurseries, Inc.
Stevensville, Michigan 49127

Farmer Seed & Nursery Company
Faribault, Minnesota 55021

Earl Ferris Nursery
Hampton, Iowa 50441

Henry Field Seed & Nursery Co.
Shenandoah, Iowa 51601

Gurney Seed & Nursery Co.
Yankton, South Dakota 57078

Hillemeyer Nurseries
Lexington, Kentucky 40500

Inter-State Nurseries, Inc.
Hamburg, Iowa 51640

Kelly Bros. Nurseries
Dansville, New York 14437

J. E. Miller Nurseries
Canandaigua, New York 14424

Monroe Nursery Company
Monroe, Michigan 48161

New York State Fruit Testing Cooperative Association
Geneva, New York 14456

Ozark Nursery
Tahlequah, Oklahoma 74464

R. H. Shumway
Rockford, Illinois 61100

Stark Bros. Nurseries
Louisiana, Mo. 63353

There are many others, some small and selling locally or just within their state those varieties that perform best in their own area. In general, the firms that have been growing and selling fruit and berry bushes nationally for many years can provide a wider choice of varieties to suit your needs. Local garden centers, nurseries, and farm-supply stores also have stock available in season.

COMMON THREATS TO FRUITS AND BERRIES

Tens of thousands of insect species inhabit this green growing world with us. Fortunately, many of them do no real harm. Only a nasty few really give us problems in our gardening and fruit growing.

Here's a list of the most troublesome for bush fruits and tree fruits. Insect guides and literature from pesticide manufacturers provide close-up details about these insect enemies so you can identify them and learn how to eliminate them.

MOST COMMON INSECTS ATTACKING SMALL FRUIT

Blackberry	aphid, cutworm, Japanese beetle, cane borer, gall caused by insects
Blueberry	blueberry maggot, Japanese beetle, gall
Currant	currant aphid, currant worm, gooseberry caterpillar
Gooseberry	same as for currant
Grapes	leafhopper, Japanese beetle, rose chafer, grape curculio, leaf tier, mealybug
Raspberry	raspberry fruitworm, cane borer, rose chafer, white grub, red spider, grasshopper
Strawberry	cane borer, cutworm, crown borer, sawfly, strawberry weevil, curculio, spettlebug

INSECTS ATTACKING SMALL FRUIT TREES

Apple	coddling moth, curculio, tent caterpillar, cankerworm, apple maggot, European red mite
Apricot	same as for peach
Cherry	plum curculio, tent caterpillar, cherry maggot
Peach	plum curculio, peach borer, Oriental fruit moth
Pear	plum curculio, pear psylla, coddling moth
Plum	plum curculio
Quince	coddling moth

FOR MORE INFORMATION ABOUT FRUIT GROWING . . .

Your best bet for additional information about fruit growing in specific areas is the agricultural extension office in states known for major fruit-producing regions. That's logical.

Here are the major states in fruit growing. Address your inquiries for bulletins to the Agricultural Extension Service Director. If you live elsewhere, you can expect to pay a slight charge for mailing to an out-of-state resident.

- New York State College of Agriculture and Life Sciences, Ithaca, New York 14850
- Rutgers University, The State University of New Jersey, New Brunswick, New Jersey 08903
- Michigan State University, East Lansing, Michigan 48823
- University of Illinois, Urbana, Illinois 61801
- University of Massachusetts, Amherst, Massachusetts 02002
- Ohio State University, Columbus, Ohio 43210
- New York State Agricultural Experiment Station, Geneva, New York 14456

And, of course, check with your own state extension service. Missouri, California, and Georgia all have exceptional literature on specific aspects of home growing. (See following list for addresses.)

WHERE TO GET HELP IN YOUR STATE

Every state has a land-grant agricultural college as part of its state university system. At these colleges you can consult a variety of horticultural specialists, from landscape experts to pomologists who serve the citizens of their state. In addition, the Cooperative Extension Service of each state is a federally and state funded organization that can be highly valuable to you as you plan and plant your fruitful landscape.

The State Extension Specialists are charged with the responsibility of providing information to homeowners and citizens of their states. In addition, each county in the United States has County Extension Specialists. These people also are well trained in providing detailed information about fruit trees, berry bushes and plants, home ground planting, and horticulture in general. Since space does not permit listing all the county extension agents, I'm providing the address of the Information Specialist at each of the land-grant colleges and universities. From that source you can get the names and local addresses of the county specialists. In general, the county specialists are at the county seat of government in each of the thousands of counties across America. In effect, an army of talented and dedicated people is available to assist you in many ways, from soil testing to advising on the best varieties of trees and shrubs, bushes, and plants that will perform best under your own local climate, soil, and other growing conditions.

Agricultural Information
Auburn University
Auburn, Alabama 36830

Agricultural Information
University of Alaska
College, Alaska 99701

Agricultural Information
College of Agriculture
University of Arizona
Tucson, Arizona 85721

Agricultural Information
University of Arkansas
Box 391
Little Rock, Arkansas 72203

Agricultural Information
Agricultural Extension Service
2200 University Avenue
Berkeley, California 94720

Agricultural Information
Colorado State University
Fort Collins, Colorado 80521

Agricultural Information
College of Agriculture
University of Connecticut
Storrs, Connecticut 06268

Agricultural Information
College of Agricultural Sciences
University of Delaware
Newark, Delaware 19711

Agricultural Information
University of Florida
217 Rolfs Hall
Gainesville, Florida 32601

Agricultural Information
College of Agriculture
University of Georgia
Athens, Georgia 30602

Agricultural Information
University of Hawaii
2500 Dole Street
Honolulu, Hawaii 96822

Agricultural Information
College of Agriculture
University of Idaho
Moscow, Idaho 83843

Agricultural Information
College of Agriculture
University of Illinois
Urbana, Illinois 61801

Agricultural Information
Agricultural Administration
Building
Purdue University
Lafayette, Indiana 47907

Agricultural Information
Iowa State University
Ames, Iowa 50010

Agricultural Information
Kansas State University
Manhattan, Kansas 66502

Agricultural Information
College of Agriculture
University of Kentucky
Lexington, Kentucky 40506

Agricultural Information
Louisiana State University
Knapp Hall
University Station
Baton Rouge, Louisiana 70803

Agricultural Information
Department of Public Information
University of Maine
Orono, Maine 04473

Agricultural Information
University of Maryland
Agricultural Division
College Park, Maryland 20742

Agricultural Information
Stockbridge Hall
University of Massachusetts
Amherst, Massachusetts 01002

Agricultural Information
Department of Information
Services
109 Agricultural Hall
East Lansing, Michigan 48823

Department of Information
Institute of Agriculture
University of Minnesota
St. Paul, Minnesota 55101

Agricultural Information
Mississippi State University
State College, Mississippi 39762

Agricultural Information
1-98 Agricultural Building
University of Missouri
Columbia, Missouri 65201

Office of Information
Montana State University
Bozeman, Montana 59715

Department of Information
College of Agriculture
University of Nebraska
Lincoln, Nebraska 68503

Agricultural Communications
Service
University of Nevada
Reno, Nevada 89507

Agricultural Information
Schofield Hall
University of New Hampshire
Durham, New Hampshire 03824

Agricultural Information
College of Agriculture
Rutgers, The State University
New Brunswick, New Jersey
08903

Agricultural Information
Drawer 3A1
New Mexico State University
Las Cruces, New Mexico 88001

Agricultural Information
State College of Agriculture
Cornell University
Ithaca, New York 14850

Agricultural Information
North Carolina State University
State College Station
Raleigh, North Carolina 27607

Agricultural Information
North Dakota State University
State University Station
Fargo, North Dakota 58102

Cooperative Extension Service
The Ohio State University
2120 Fyffe Road
Columbus, Ohio 43210

Agricultural Information
Oklahoma State University
Stillwater, Oklahoma 74074

Agricultural Information
Oregon State University
Carvallis, Oregon 97331

Agricultural Information
The Pennsylvania State
University
Room 1, Armsby Building

University Park, Pennsylvania 16802

Cooperative Extension Service
University of Puerto Rico
Mayaguez Campus, Box AR
Rio Piedras, Puerto Rico 00928

Agricultural Information
University of Rhode Island
16 Woodwall Hall
Kingston, Rhode Island 02881

Agricultural Information
Clemson University
Clemson, South Carolina 29631

Agricultural Information
South Dakota State University
University Station
Brookings, South Dakota 57006

Agricultural Information
University of Tennessee
Box 1071
Knoxville, Tennessee 37901

Department of Agricultural Information
Texas A & M University
College Station, Texas 77843

Agricultural Information
Utah State University
Logan, Utah 84321

Agricultural Information
University of Vermont
Burlington, Vermont 05401

Agricultural Information
Virginia Polytechnic Institute
Blacksburg, Virginia 24061

Agricultural Information
115 Wilson Hall
Washington State University
Pullman, Washington 99163

Agricultural Information
Evansdale Campus
Appalachian Center
West Virginia University
Morgantown, West Virginia 26506

Agricultural Information
University of Wisconsin
Madison, Wisconsin 53706

Agricultural Information
University of Wyoming
Box 3354
Laramie, Wyoming 82070

Information Services
Federal Extension Service
U.S. Department of Agriculture
Washington, D.C. 20250

Index